Chromosome hierarchy

BERNARD JOHN
and
KENNETH R. LEWIS

Chromosome Hierarchy

AN INTRODUCTION
TO THE BIOLOGY
OF THE CHROMOSOME

1975 · CLARENDON PRESS · OXFORD

Oxford University Press, Ely House, London W. 1

GLASGOW NEW YORK TORONTO MELBOURNE WELLINGTON
CAPE TOWN IBADAN NAIROBI DAR ES SALAAM LUSAKA ADDIS ABABA
DELHI BOMBAY CALCUTTA MADRAS KARACHI LAHORE DACCA
KUALA LUMPUR SINGAPORE HONG KONG TOKYO

Casebound ISBN 0 19 857160 7
Paperback ISBN 0 19 857161 5

© Oxford University Press 1975

Printed in Great Britain
by Fletcher & Son Ltd., Norwich

Preface

Self-replication, the diagnostic feature of the genetic material, can
proceed properly only in a system which is designed to mediate between
it and the supporting environment. An organism is such a system but one
which is itself a product of interactions between the genotype and the
environment. Indeed, the development of the organism, which allows
the perpetuation of the genotype, itself depends on the perpetuation of
the genotype. It also depends on the capacity of the genetic material to
specify molecules which are not of its own kind.

A third feature of the genetic material is its capacity for change,
ultimately by mutation and more immediately by recombination. These
too are essential for perpetuation in the long term because they create
the diversity which allows the exploitation of environmental variety.

For the most part, the genetic material is contained in the cell nucleus
and is organized into chromosomes which are in many ways witness to
their own industry. Thus, the extent to which they recombine at meiosis
is expressed in the number of visible chiasmata they form, while selective
gene activity in certain cells is revealed in reversible localized changes in
chromosome morphology. Structural mutations, on the other hand, are
reflected in stable alterations in appearance and pairing behaviour.

However, both the genetic diversity and the epigenetic modifications
of chromosomes are largely concealed by a uniformity of basic structure
and behaviour during division. In this book, we begin with the essentially
uniform characteristics of chromosomes before considering their contribution to development and their involvement in evolution.

B.J.
K.R.L.

A deck of cards was built like the purest
of hierarchies, with every card a master to
those below it, a lackey to those above it.
And there were "masses"–long suits—which
always asserted themselves in the end,
triumphing over the kings and aces.

Ely Culbertson (1943). *Total peace*. Faber, London.

Contents

1 *Introduction—nucleus and chromosome*

Biological systems, like physico-chemical systems, are atomic in structure and design. The atoms from which biological systems are constructed are neither unique nor even rare. About three-quarters of them are oxygen; hydrogen and carbon each contribute about 10 per cent while nitrogen comprises a little over 2 per cent. These four atoms are among the lightest and most active known and occur abundantly in air, soil, and water. They are also adaptable and associative and readily can be assembled, disassembled, and re-assembled according to various structural plans.

In living systems, atoms have a very precise, orderly arrangement at molecular and supra-molecular levels. The molecules include water (the principal component), inorganic salts, and organic compound in the form of carbohydrates, lipids, proteins, and nucleic acids. While life processes cannot proceed without water and minerals, they take their most distinctive properties from the organic molecules. The reason for this is not difficult to appreciate. Order and disorder at the biochemical level are essentially problems of organization, and since order calls for a particular state of organization it can arise only from previous order. And the order of biological systems stems predominantly from two classes of macromolecules—the nucleic acids and the proteins. Both of these are polymers, containing repetitive linear aggregates of unit monomers. In proteins the monomer units are x-amino acids, about twenty of which occur naturally, while in nucleic acids they are nucleotides (Fig. 1).

Nucleotides consist of a combination of a heterocyclic organic base (a purine or pyrimidine) with a particular sugar residue and a phosphoric acid residue (base + sugar = nucleoside; nucleoside + phosphate = nucleotide). In ribonucleic acid (RNA) the sugar concerned is ribose while in deoxyribonucleic acid (DNA) deoxyribose sugar is involved. The purines, adenine (A) and guanine (G), and the pyrimidine, cytosine (C), are common to both these acids. But the thymine (T) of DNA is

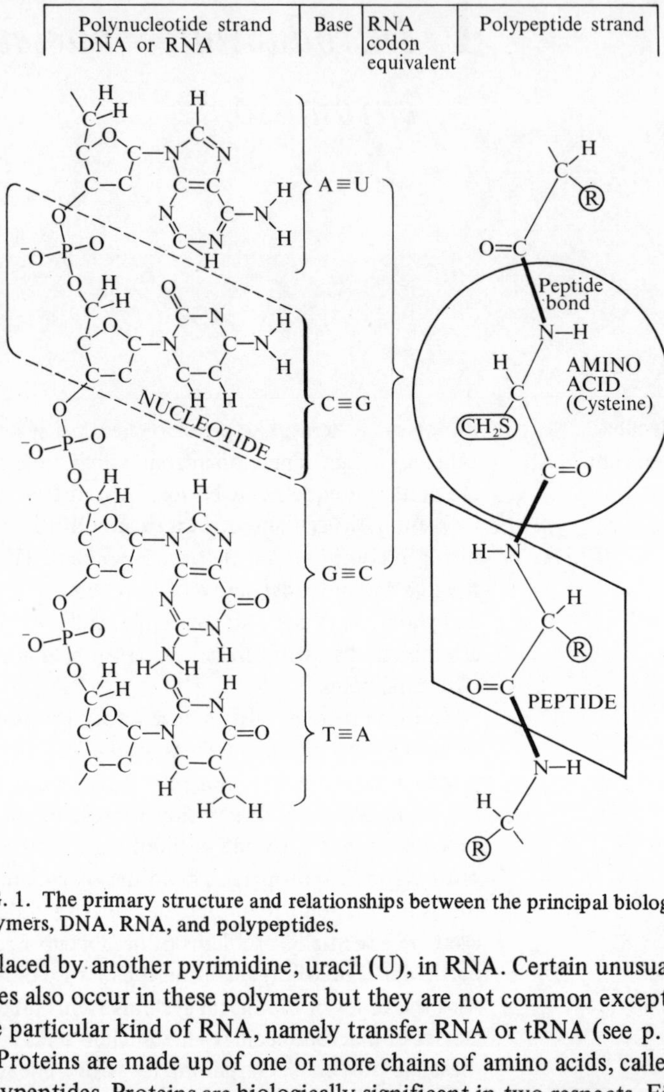

FIG. 1. The primary structure and relationships between the principal biological polymers, DNA, RNA, and polypeptides.

replaced by another pyrimidine, uracil (U), in RNA. Certain unusual bases also occur in these polymers but they are not common except in one particular kind of RNA, namely transfer RNA or tRNA (see p. 3).

Proteins are made up of one or more chains of amino acids, called polypeptides. Proteins are biologically significant in two respects. First, they form a molecular skeleton which governs biochemical topology, for they guide atoms and molecules into precise locations as a result of the arrangement of their own atoms. To this end structural proteins are organized into systems of particles, fibres, and membranes. Second, some proteins have the special property of speeding up the rates of specific biochemical reactions. Without these biocatalysts, or enzymes, all life processes would cease, for all macromolecules are synthesized through the mediation of specific enzymes.

Proteins themselves, however, can be synthesized only in the presence of nucleic acids, certain forms of which act as molecular mid-wives during the birth of the new protein molecule. The sequence of nucleotides, the only variable feature in the primary structure of the nucleic acid polymer, in fact specifies the sequence of α-amino acid monomers in particular polypeptides. The sequence of nucleotides which determines the amino acid sequence in a polypeptide chain is known as a cistron. Nucleic acids are significant also because they represent the only known molecular species with a capacity for self-replication. It is, therefore, the nucleic acid component of an organism that governs the storage, readout, and transfer of genetic information.

The order of biological systems is thus determined by, and developed from, pre-coded information. The coding system is molecular and the primary information molecule is nucleic acid. Reproduction depends on the replication of this informational system and its transmission from parent to offspring. Development, on the other hand, involves the nucleic-acid-dependent synthesis of those specific structural and enzymic proteins which are necessary for the organization of the new individual. And evolution depends ultimately on changes in the informational system, changes which lead to modified patterns of protein structure and, hence, to altered developmental and metabolic pathways.

In all organisms, therefore, a nucleic acid component represents the fundamental genetic material and fulfils two basic functions. On the one hand, it serves as a template for its own replication (autocatalytic function): this requires a suitable supply of precursors, enzymes, and energy, and the nucleic acid molecule itself must be in the correct configurational state. On the other hand, it also supplies the structural and regulatory information for the synthesis of specific proteins (heterocatalytic function).

In all cellular organisms, and those viruses in which deoxyribonucleic acid (DNA) serves a genetic function, its heterocatalytic role depends upon the formation of an intermediary messenger molecule of ribonucleic acid (mRNA), the product of a process termed transcription, which then functions as a secondary template for protein synthesis, in a process called translation. This translation process always take place on specialized organelles called ribosomes, which serve a central role in protein biosynthesis. These ribosomes are ribonucleoprotein particles, the RNA of which (rRNA) is also templated by DNA in the nucleus, called rDNA. Each messenger molecule consists of a sequence of specific codons, each comprising a triplet of bases that codes for a single amino acid in the polypeptide which is constructed at translation (Table 1). Special RNA molecules, transfer RNA (tRNA), each containing a triplet anticodon complementary to a particular codon on the mRNA strand, then serve to bring particular amino acids into correct alignment at the ribosome site (Fig. 2).

Table 1

The codons of messenger RNA and the amino acids they specify

5' terminal	Middle nucleotide				3' terminal
	U	C	A	G	
U	UUU } Phe UUC UUA } Leu UUG	UCU UCC } Ser UCA UCG	UAU } Tyr UAC UAA } Term UAG	UGU } Cys UGC UGA Term UGG Trp	U C A G
C	CUU CUC } Leu CUA CUG	CCU CCC } Pro CCA CCG	CAU } His CAC CAA } Gln CAG	CGU CGC } Arg CGA CGG	U C A G
A	AUU AUC } Ile AUA AUG Met†	ACU ACC } Thr ACA ACG	AAU } Asn AAC AAA } Lys AAG	AGU } Ser AGC AGA } Arg AGG	U C A G
G	GUU GUC } Val GUA GUG	GCU GCC } Ala GCA GCG	GAU } Asp GAC GAA } Glu GAG	GGU GGC } Gly GGA GGG	U C A G

Notes. Term = known terminating codons in *E. coli,* cannot be recognized by any known tRNA
† Serves as an initiation codon in *E. coli*
Abbreviations
Bases. A—adenine, C—cystosine, G—guanine, U—uracil
Amino acids. Ala—alanine, Arg—arginine, Asn—asparagine, Asp—aspartic acid, Cys—cysteine, Glu—glutamic acid, Gln—glutamine, Gly—glycine, His—histidine, Ile—isoleucine, Leu—leucine, Lys—lysine, Met—methionine, Phe—phenylalanine, Pro—proline, Ser—serine, Thr—threonine, Trp—tryptophan, Tyr—tyrosine, Val—valine.

In some cases, the mRNAs corresponding to two or more adjacent cistrons of the DNA are synthesized sequentially so that they become linked into one long, polycistronic mRNA which may remain intact during the process of translation. In certain viruses the genetic material itself consists of RNA which can function directly, in the manner of mRNA, as a template for polypeptide production.

The synthesis of proteins is the initial step in all production sequences and, therefore, in the epigenetic expression of all hereditary information as well (Fig. 3). Thus genetic material must be able to contain and maintain information, information which can be duplicated and transmitted during reproduction and transcribed and translated during development. From a functional point of view, that segment of DNA which is ultimately responsible for specifying the primary structure (sequence of amino acids) of a particular type of polypeptide chain can be regarded as a fundamental epigenetic unit, a cistron, or gene (Table 2). Errors of

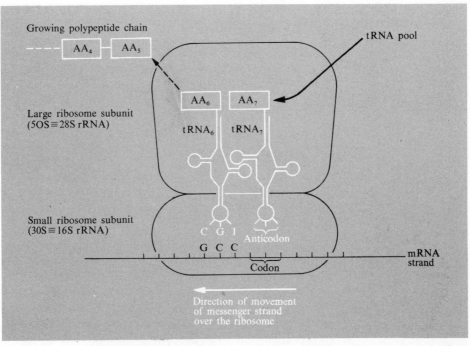

FIG. 2. The relationship between ribosomes, tRNA, mRNA, and the production of a polypeptide strand.

Table 2

Components of the hereditary hierarchy

Unit of DNA	Molecular length	No. base pairs	No. of combinations	Character
1. Base pair	3·4 Å	4	4	Ultimate unit of mutation
2. Codon equivalent	10 Å	3	64	Determines single amino acid
3. Cistron	0·3 nm	average 1000	10^{200}	Determines single polypeptide chain
4. Operon	1 nm	1000–5000		Unit of co-ordinate transcription
5. Chromosome (man)	4 cm	$1·3 \times 10^9$	$10^{3\ 000\ 000}$	Unit of independent assortment
6. Genome (man)	1 m	3×10^9	$10^{10\ 000\ 000}$	Unit of genetic control

Note. After Papazian (1967).

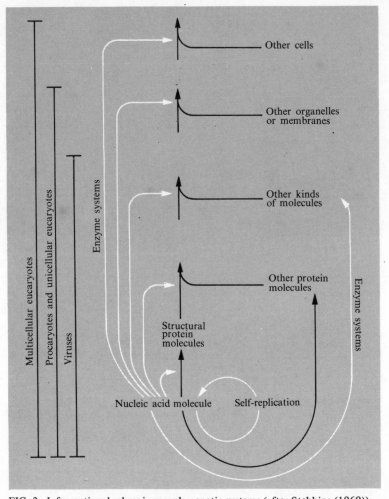

FIG. 3. Informational relays in morphogenetic systems (after Stebbins (1969)).

replication which involve permanent changes in the coding system and which lead to modifications in the transcription-translation mechanism give rise to variant polypeptides. Such errors are termed mutations and they can involve the insertion, deletion, or substitution of nucleotides or changes in nucleotide sequence.

Gene mutations can be considered in two main groups according to the nature of the gene in which they occur and, hence, the process by which they exert their effects. First, those which occur in structural cistrons, which are responsible for mRNA production, and, secondly, those which arise in tRNA-coding cistrons. The former will affect specific polypeptides whilst the latter are expected to have a wider sphere of influence.

1.1.1. *Mutations in structural cistrons*

(a) *Missense mutations.* A base substitution affecting a structural cistron usually results in an amino-acid substitution at a specific position in a polypeptide chain. Such mutations are termed missense. Base locations in the genes are collinear with those of the amino acids in the polypeptide they specify. Consequently, the site of amino-acid substitution corresponds with that of the missense mutation. The nature of the substitution will depend, of course, on the precise nature of the base change. And a knowledge of the genetic code allows the base change to be specified, with some precision, from the substitution it produces. Some types of amino acid substitution are expected to be of little consequence. For example, the replacement of glycine by alanine, valine by leucine, or phenylalanine by tyrosine. Such substitutions are termed conservative. On the other hand, the replacement of leucine (neutral) by arginine (basic) or aspartic acid (acidic) by valine (neutral) are more drastic since they change the net molecular charge of the chain. The replacement of cysteine by another amino acid is particularly disastrous to the functional performance of the polypeptide, since cysteine is unique among the amino acids in having an SH group that is often essential to the activity of an enzyme. Likewise intra-chain disulphide bridges formed between pairs of cysteines frequently determine the molecular shape of a polypeptide (tertiary structure), or else the development of quaternary structure in the case of polymeric molecules. The OH group of serine and histidine also serves as an active radical for certain enzymes so that their replacement too can have drastic consequences.

(b) *Samesense mutations.* All the amino acids except methionine and tryptophan can be specified by more than one codon. Indeed, in the case of glycine, alanine, valine, threonine, and proline the third base of the codon is redundant. With isoleucine, phenylalanine, tyrosine, aspartic acid, asparagine, glutamic acid, glutamine, and histidine, on the other hand, the third base of the codon is only partly redundant. For example, while the first two bases of the codon for tyrosine read UA, the third can be either U or C but not A or G. In these cases of total or partial redundancy the substitution of the third base does not necessarily alter the amino acid sequence of the polypeptide. For this reason such base substitutions are termed samesense mutations.

(c) *Nonsense mutations.* If a base substitution changes an amino-acid-specifying codon to a chain-terminating codon it will lead to premature chain termination. Such mutations are termed nonsense.

(d) *Frameshift mutations.* The triplet codons of mRNA are not defined by a system of punctuation marks. The bases are simply translated sequentially in groups of three from a fixed starting point at which amino acid assembly begins. If, therefore, any *three* adjacent bases, which form a codon according to the 'reading frame' described above, are deleted, the resulting polypeptide will simply lack an amino acid at a location which corresponds to that of the site of deletion.

However, the deletion of one single base, or two adjacent bases, 'smaller' mutations in the absolute sense, will have much more drastic consequences. This follows from the fact that the bases are read sequentially in threes from a fixed initiation site according to the sequence in which they lie. Consequently, the deletion of a base will alter not only the content of the codon in which it previously occurred, but that of all other codons lying distal to the initiation point which are scheduled for later translation. This reframing of codons can lead to various combinations of missense, samesense, and nonsense change.

1.1.2. Mutations affecting tRNA genes	An organism has over 60 kinds of translatable codons in its mRNA but only 30 or so types of tRNA. This is because the number of anti-codons required for recognizing particular synonymous codons can be reduced by introducing unusual bases at the third position of the anti-codon. For example, hypoxanthine (HyX) is a derivative of A but unlike A it can pair not only with U but also with C and even A. Thus an alanine tRNA having the anticodon CG HyX can recognize three of the four codons for alanine, namely, GCU, GCC, and GCA.

Since a large variety of tRNA species is likely to be required for the translation of each mRNA sequence, and since many kinds of tRNA are involved with many different mRNA molecules, mutations affecting tRNA cistrons are likely to have widespread effects. Two rather different types of tRNA mutations are known.

(a) *Suppressor mutations.* UAG, UAA, and UGA can serve as chain-terminating codons, presumably because there are no genes for tRNAs with matching anticodons. If the tyrosine tRNA anticodon AUG is mutationally altered by a single base substitution to AUC, then the mutant tRNA produced will still carry tyrosine, but will recognize UAG; and presumably then add tyrosine to the growing polypeptide. In this event UAG no longer serves as a nonsense codon indicating the termin-ation of translation. Such a tRNA mutation can, therefore, suppress the expression of interstitial nonsense mutations in genes which specify mRNA because they allow translation to proceed beyond the stop sign.

(b) *Mutations resulting in ambiguous coding.* Normally the codon AAG on mRNA is recognized only by lysine tRNA. If a mutation in the glutamine tRNA cistron were to occur, changing its anticodon from GUC to UUC, then the mutant glutamine tRNA would also now recognize the AAG codon. Such a mutation is likely to cause ambiguous coding in the sense that if an mRNA contains an AAG codon then a single structural cistron in the genome will produce, in the presence of the above altered tRNA, two or more polypeptide chains, differing from one another, as one or more will have glutamine in the place of lysine.

While mutations affecting tRNA cistrons, together with frameshift and nonsense mutations, must be almost invariably deleterious, same-sense and most of the conservative missense mutations are neutral, or nearly so.

As a consequence of mutation, a given gene locus may exist in two or more so-called allelic states. Allelic genes influence the same component of development but do so in different ways or to different extents and so lead to the production of variant states.

1.2. Macromolecular organization

The simplest macromolecular complex that has sufficient information for its own replication and reproduction is the virus. Viruses are intracellular parasites which are unable to metabolize in a free state, so that outside their specific host cells they are biologically inert. A virus particle usually consists of one large molecule of nucleic acid, ranging in molecular weight from 1–200 million, which is enveloped in a protective coat of protein, the capsid. The coat furnishes a suitable transportive covering for the infective principle, the nucleic acid, the functions of which are the storage of genetic information and its transfer to progeny particles. At infection, the nucleic acid is released from the capsid into an appropriate host cell within which it initiates changes in metabolic pattern which lead eventually to the development of new viral nucleic acid and specific viral capsid protein.

Viruses are of various biochemical types. Some animal viruses (e.g. poliomyletis), the bacterial viruses f_2 and R17, MS2, and Qβ, and most plant viruses contain only RNA. For example, the RNA of Qβ consists of only about 3500 nucleotides and comprises only three known cistrons which code for the assembly or maturation protein (A_2 protein), which is a structural component of the viral particle; the coat protein; and the B-subunit of an enzyme (replicase) required for viral replication. Other viruses and indeed all other organisms, have DNA as their genetic component, while cellular forms have also non-genetic or metabolic RNA. In most organisms the DNA is dimeric, consisting of a double helix of two polynucleotide columns. In the bacterial viruses ϕX174, ϕR, and S13, however, while the DNA molecules formed during the vegetative phase of the virus are duplex, genetic information is transferred to progeny virus in the form of single polynucleotide sequences. But in all DNA viruses, regardless of their precise type, the nucleic acid component appears to be organized as a single molecule. This is usually tightly packed and folded within the protein capsid but it assumes a filamentous form when released from the confines of the coat either naturally, as at infection, or artificially by osmotic or detergent treatment. The processes of nucleic acid replication and transcription thus occur when the molecule is in an extended state.

In contrast to viruses, all bacteria and blue-green algae (procaryota) and all plants and animals (eucaryota) are made up of one or more cells. There are, however, distinct differences in the organization of pro- and eucaryotic cells with regard to both the 'cytoplasm' and the 'nucleus' (Table 3). Particularly striking is the absence of a complex cytoplasmic architecture in procaryotes. This is not difficult to understand when one

Table 3
Cell organization in procaryotes and eucaryotes

		Cell type	
	Component	Procaryotic	Eucaryotic
Nuclear	Membrane	Absent	Present
	Nucleolus	Absent	Present
	DNA	Naked duplex	Combined with histone
	Linkage groups	Single and circular	Multiple and linear
	Mitosis and meiosis	Absent	Present
Cytoplasmic	Spindle	Absent	Present
	Ribosomes	70S (30S + 50S)	80S (60S + 40S)
	Enzymes of respiratory and photosynthetic electron transport system	Localized on cell membrane	Localized on specialized membranous organelles (mitochondria and chloroplasts)
	Photosynthetic organelles	Simple chromophores	Complex chloroplasts
	Cell wall	Contains muramic acid	Never contains muramic acid

considers that such cells are self-sufficient as organisms. They do not require to communicate with other cells. In addition they have to respond rapidly to environmental alterations.

As far as the DNA component of procaryotes is concerned, it is concentrated in a particular area of the cell to form a nucleoid which is not limited by a distinct membrane. Its shape is variable, ranging from a delicate network of branches to a compact nuclear body. Nucleoids can be disrupted at an air-water interface and the liberated DNA can be incorporated into a monomolecular film of suitable basic protein. These films can be picked up on grids, shadowed with platinum C, and examined with the electron microscope. The most conspicuous feature in such preparations is the apparent absence of ends or ramifications in the nucleic-acid molecule, which indicates that the entire DNA pool is in fact a single unbranched chain. From autoradiographic examination of the silver grain contours produced by tritium (^{3}H) decay in such isolated and labelled DNA fibres, it has been estimated that the DNA molecules in vegetative cells of *E. coli* are 100–400 μm long, and circular. Likewise the DNA of *Mycoplasma hominis* is also a ring approximately 265 μm long. It is not clear whether bacterial DNA in general is organized in an equivalent manner, although this is usually assumed to be the case. It must be emphasized that this conception of a ring-shaped bacterial genome has no simple morphological counterpart in terms of the electron-microscope structure of the intact nucleiod. Since the DNA filament is about 1000 times longer than the cellular space that houses it, the DNA fibre must be considerably folded within the nucleoid, though how this is achieved is not known.

The genetic mechanism of eucaryotes differs from that of procaryotes in three important respects.

(a) The DNA of eucaryotes is generally complexed with basic protein (histone) to form a system of microscopic fibres known as chromosomes (*chrom-* = colour, *soma* = body). These represent cell organelles specialized for the storage, replication, and transmission of genetic material both within and between generations. According to its replicative state, and hence the amount of DNA it contains, a chromosome may consist of one chromatid or two chromatids, or in the special case of polytene chromosomes, which we shall meet later (see p, 35). 2^n chromatids where n can be as high as eleven.

(b) There is usually considerably more DNA per nucleus in eucaryotes. This increase is, of course, compatible with the greater amount of information required for the development of eucaryotes, which are more complex. Even so, all the increase cannot be accommodated in these terms.

If aqueous solutions of duplex DNA are raised to high temperatures, the two polynucleotide strands of the dimer separate. Slow cooling of such denatured dimers leads to renaturation. By renaturing mixtures of heat-denatured DNA from different sources it is possible to produce hybrid molecules. This technique provides a means of assessing the extent of base-sequence correspondence or homology between different DNAs or indeed between DNA and RNA, since the rate of renaturation is a function of the concentration of the complementary base sequences represented in the mixture.

Molecular hybridization studies of this kind have shown that large portions of the genome in all eucaryotes are unique in character. Even so, clustered repeat nucleotide sequences have been found in the DNA of nearly all the eucaryotes so far studied. This includes such simple creatures as the flagellate *Euglena*, the sponge *Microciona*, and the tunicate *Ciona*. These repeat sequences are of two kinds:

(*i*) a fast reassociating fraction containing in excess of 10^5 copies of a given sequence per haploid genome, and

(*ii*) an intermediate fraction with some 10^2-10^5 copies per haploid genome.

When DNA is analysed by caesium chloride (CsCl) or caesium sulphate (Cs_2SO_4) density gradient centrifugation, the highly repetitious fraction can often be shown to be associated with a minor or satellite peak which is quite distinct from the main band. A satellite DNA which is enriched in G and C will show a higher bouyant density, while one enriched in A and T will show a lower bouyant density than main band DNA. Some species have 'hidden satellites' which have the same mean base composition as main band DNA but still contain clustered repetitious sequences. Such hidden satellites can be revealed if the DNA is centrifuged with $CsSO_4$ to which Ag^+ or Hg^{++} has been added. These metallic ions bind to particular bases, making the satellites significantly lighter or heavier than main band DNA.

The kinetics of renaturation indicate that no gross repetition occurs in procaryotes though a small amount of repeat DNA is present in the genes coding for all three classes of ribosomal RNA (5S, 16S, and 23S). This, however, is several orders of magnitude lower than the degree of repetition found in eucaryotes. Thus in *Xenopus laevis* there are about 450 copies per haploid complement of the genes coding for 18S and 28S rRNA and not less than 9000 copies of the gene for 5S rRNA (Table 4). Likewise in *Drosophila,* where there are 130 copies of the 18S and 28S genes in both the X- and the Y- chromosome, 13–20 copies are present of each of the 30–40 different kinds of tRNA.

Table 4

Redundancy of ribosomal RNA genes in biological systems

Organism	Percentage of genome homologous to rRNA	Daltons per nucleus		Number of rRNA genes per nucleus
		Total DNA	rDNA	
E. Coli	0·30	$4·0 \times 10^9$	$1·2 \times 10^7$	6
Chicken	0·30	$1·4 \times 10^{12}$	$3·6 \times 10^9$	200
D. melanogaster	0·27	$2·4 \times 10^{11}$	$6·4 \times 10^8$	260
Human HeLa cells	0·008	$1·2 \times 10^{13}$	$9·6 \times 10^8$	440
Xenopus laevis	0·1	$3·6 \times 10^{12}$	$3·6 \times 10^9$	1600

Note. One Dalton is the mass of one hydrogen atom

The presence of repetitious DNA goes some way towards explaining the increased DNA content of eucaryotes. Even so the proportion of the genome that is repeated varies considerably in different eucaryotes, even between closely related species. This applies particularly to the satellite fraction where large differences occur both in quantity and composition (see p. 23).

(c) In both pro- and eucaryotes the units of DNA which code for the various proteins and non-translated RNA molecules of the cell are integrated in a specific order into larger units or linkage groups— chromosomes. These groups not only maintain and replicate the genetic information contained in their DNA but, in addition, ensure that this information is transcribed at the right time and in the proper sequence into the specific RNA types which direct protein synthesis. In pro- caryotes there is only one linkage group but in eucaryotes there are at least two and usually many more chromosomes per genome (Table 5).

Table 5

Indication of the range of chromosome numbers found in eucaryotes

Plants	Diploid chromosome number	Animals
Haplopappus gracilis	4	*Parascaris equorum* var. *bivalens* (horse threadworm)
Luzula purpurea (woodrush)	6	*Aedes aegypti* (yellow fever mosquito)
Crepis capillaris	8	*Drosophila melanogaster* (fruitfly)
Vicia faba (field bean)	12	*Musca domestica* (house fly)
Brassica oleracea (cabbage)	18	*Chorthippus parallelus* (grasshopper)
Citrullus vulgaris (water melon)	22	*Cricetulus griseus* (Chinese hamster)
Lilium regale (royal lily)	24	*Schistocerca gregaria* (desert locust)
Bromus texensis	28	*Desmodus rotundus murinus* (vampire bat)
Camelia (Thea) sinensis (chinese tea)	30	*Mustela vison* (mink)
Magnolia virginiana (American sweet bay)	38	*Felis catus* (domestic cat)
Arachis hypegaea (peanut)	40	*Mus musculus* (mouse)
Coffea arabica (coffee)	44	*Mesocricetus auratus* (Syrian or golden hamster)
Stipa spartea (porcupine grass)	46	*Homo sapiens* (modern man)
Chrysoplenum alternifolium (golden saxifrage)	48	*Pan troglodytes* (chimpanzee)
Aster laevis (Michaelmas daisy)	54	*Ovis aries* (domestic sheep)
Glyceria canadensis (rattlesnake manna grass)	60	*Capra hircus* (goat)
Carya tomentosa (hickory)	64	*Dasypus novemcinctus* (nine-banded armadillo)
Magnolia cordata	76	*Ursus americanus* (American black bear)
Rhododendron keysii	78	*Canis familiaris* (dog)

Note. Much higher numbers are known in certain plant groups, e.g. pteridophytes, but these are exceptional.

The chromosomes of eucaryotes go through distinctive division cycles called mitosis and meiosis. Between these division sequences they are enclosed within a special nuclear membrane and occur predominantly as highly extended and invisible linear entities. During division these extended interphase chromosomes condense into short, thick rods (prophase) which orient within a spindle-shaped system of microtubules (metaphase) prior to dividing into two groups (anaphase) from each of which a new membrane-bound nucleus (telophase) is reconstituted. The details of the division mechanism vary to some extent with species but more importantly with the kind of division concerned (mitosis vs meiosis) and we shall deal with the variations later (see p. 118). For the moment it is sufficient to emphasize that mitosis is the nuclear division employed during development and for clonal or asexual reproduction. It leads essentially to the production of genetically identical nuclei. Meiosis, on the other hand, is associated with the means of combining two lines of heredity during reproduction and is more often associated with sexual differentiation. Sexual reproduction in fact involves a combination of two alternating and compensating processes—meiosis and fertilization. But there can be reproduction without fertilization by sexually differentiated females (parthenogenesis), and two lines of heredity can be combined and recombined even in the absence of morphologically recognizable sexual differentiation (e.g. conjugation in ciliate protozoa; fusion of fungal hyphae).

2 *Chromosome architecture*

2.1. The metabolic nucleus

2.1.1. Chemical composition

Chromosomes contain molecules of four kinds—DNA, RNA, low molecular-weight basic proteins calles histones, and more complex acidic proteins which are often referred to as residual proteins. Collectively these molecules constitute what is called chromatin. Except for the case of certain enzymes concerned with transcription or replication, the function of chromosomal proteins is not fully understood. No strictly structural proteins have been identified and the linear integrity of DNA throughout the length of the chromosome does not seem to involve protein. Indeed Kavenoff and Zimm (1973) have recently isolated DNA molecules from *Drosophila* that are long enough to contain all the DNA of a single chromosome.

DNA is the only permanently conserved molecule in the chromosome and, even during its replication, there is no replacement of the existing molecule. Instead, each of the polynucleotide strands which make up the DNA duplex serve to template a new complementary strand. Replication is therefore semi-conservative. By contrast, all the other molecular components of the chromosomes are completely replaced over successive cell cycles.

DNA and histone form a nucleoprotein complex which makes up 60–90 per cent of the bulk of a chromosome and represents the fundamental molecular basis of chromosome structure. This is consistent with the concept of DNA as the chemical basis of inheritance. Native DNA in its acid form is, however, unstable since repulsions between the negatively charged phosphate groups tend to extend the molecule and separate the component stands of the duplex. In bacteria these negative charges appear to be neutralized by the formation of complexes between DNA and either divalent ions (Mg^{++}, Ca^{++}) or else polyamines (putrescine and spermidine). In eucaryotes, on the other hand, histones appear to play an equivalent role in maintaining the stability of the DNA molecule. Some of the histones appear to be wound helically around the DNA duplex; others combine periodically with these and with the phosphate groups of the DNA backbone.

This association between histone and nucleic acid appears to have
nothing to do with the manner in which genetic information is carried
or with the basic mechanisms of replication and transcription, which
are inherent properties of DNA. Rather it appears to be a means of
regulating these processes. In procaryotes the total genetic potential of
the genome can be expressed in any cell generation. In eucaryotes, on
the other hand, only a fraction of the total genome functions at any
one time (Protozoa) or in any one cell (Metazoa). The remainder is
either inactive or repressed, and histone plays a role in this repression
(see p. 86). This role is, however, relatively non-specific and there is
evidence that the acidic proteins include molecules which not only
antagonize the repression effects of histones but are also capable of
independently repressing transcription. The amount of residual protein
and RNA varies with the metabolic state of the nucleus and these molecules do not seem to play any integral role in maintaining chromosome
structure.

Metaphase chromosomes can now be isolated in quantities sufficient
for chemical analysis. Such analysis shows that these condensed chromosomes contain 13—16 per cent DNA, 12—13 per cent RNA, and
68—72 per cent protein. These results contrast markedly with those
obtained for isolated nucleoprotein where the ratio of DNA to histone
is approximately 1:1. The increased protein content of isolated chromosomes may stem, in part, from an association with some component
of the nuclear sap during the isolation procedure. But even when allow-
ance is made for such contamination, isolated whole chromosomes still
appear to contain an additional, non-histone, protein component that is
not represented in isolated nucleoprotein.

The possible role of contamination in respect of the high percentage
of RNA in isolated metaphase chromosomes is similarly difficult to
evaluate. The presence of RNA is to be expected in cells engaged in active
protein synthesis, and it is now known that chromosomes in interphase
and in early and mid-prophase are involved in RNA synthesis. At least
some of the RNA found in chromosomes is thus likely to represent a
transitory complex synthesized by the DNA, rather than a structural
molecule. The transient nature of the association between RNA and
chromosomes is underscored by the complete absence of any RNA in
the nucleus of late stages of spermiogenesis and in the mature sperm
head.

Finally, several studies have implicated the divalent ions Ca^{++} and Mg^{++}
as necessary elements in the maintenance of chromosome structure. The
basis for such a conclusion rests on the fact that these ions are necessary
for preserving the light microscope morphology of isolated chromosomes,
while their absence has an effect on the state of chromosome coiling. It
is not clear, however, whether these ions have a structural role in the

intact chromosome or are merely necessary during fixation or isolation to maintain or replace an ionic environment disturbed by these procedures.

In summary, the precise mode of binding of protein and RNA to DNA and the molecular interactions of the non-genetic chromosomal components in replication, RNA-synthesis, coiling, and recombination are largely unknown.

2.1.2. Physical topography

Current opinion assumes that the basic unit of chromosome organization is a deoxyribonucleoprotein (DNP) fibre. That chromatin has a fibrous organization is confirmed by electron-microscope studies, but such fibres are not visible in light-microscope preparations. With conventional electron-microscope preparations the fibres have a width of 100 Å and it assumed that these consist of a 20 Å double helix of DNA combined into a 30 Å uncoiled DNP fibre, which by supercoiling makes a fibre of 100 Å. If chromosomes are spread on a distilled-water surface and then picked up on electron microscope grids, the fibres seen range in width from 200 Å to 400 Å with a mean of 250 Å. Whether the 250 Å fibril results from additional supercoiling of the 100 Å unit or whether it arises from two 100 Å fibres folded on themselves is not clear. According to DuPraw (1970) and Comings (1972) chromosomes consist of a single 100 Å fibril which loops and folds on itself sequentially (Fig. 4).

The various molecules present in the chromosome are not uniformly distributed along its length. Even in their condensed state, chromosomes can be seen to be functionally differentiated into regions of at least three kinds, each of which reflects a distinctive biochemical organization. These are (a) the centromere or kinetochore; (b) the nucleolar organizer; and (c) heterochromatic sections. Let us see what they involve.

(a) *The centromere.* Whereas the chromosomes of interphase nuclei are predominantly diffuse and extended, those of dividing cells are compact structures. This compaction depends on a process of chromosome coiling. The mechanics of coiling are unknown, though histone proteins appear to play a part in the process. In the majority of eucaryote chromosomes, coiling is differential in the sense that a particular region of each chromosome invariably refrains from coiling to the same extent as the remainder of the thread (Figs 7 and 14). This region is called the centromere or kinetochore. This organelle is a specialized genetic locus responsible, in some as yet undefined way, for the reversible binding between the chromatid and the microtubule protein of the spindle system within which the chromosomes orient and move at cell division. This function cannot simply be acquired by other regions of the chromosome, since spontaneous or experimentally produced fragments which lack a centromere (acentric fragments) fail to engage with the spindle and hence fail to move in a regular manner.

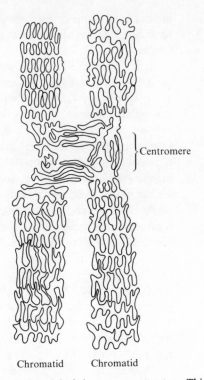

Centromere

Chromatid   Chromatid

FIG. 4. A single-stranded model of chromosome structure. This implies that a single DNP fibril folds back and forth on itself to build up one width of the chromatid and runs uninterrupted from one end of a chromatid to the other. The fibres from the two chromatids of a chromosome are assumed to interdigitate on either side of the centromere (after Comings (1972)).

As far as is known, all centromeres are functionally equivalent and, presumably, homologous entities. Thus a centromere appears to work equally well when transferred naturally or experimentally to another part of the same chromosome or even to a different chromosome. There is evidence that spindle microtubules can sometimes attach at sites other than the true centromere, but such neo-centromeres do not appear to be stable, permanent entities and in some cases, at least, their expression depends on the presence of the main centromere.

Although a majority of chromosomes have a single localized centromere, some groups of invertebrates and certain flowering plants have a quite distinct pattern of organization. Here the entire surface of the chromosome appears to be capable of eliciting spindle attachment. In neither this case, nor that of the localized centromere, however, have we any understanding of the mode of association of the chromosome with the spindle system.

Fine-structure studies have shown the localized centromere to be a complex region consisting of a plate-like outer layer set away from the chromosome surface, an inner layer which is intimately associated with, and probably composed of, chromatin fibres, with a middle layer *c.* 150–350 Å in thickness separating the two. The spindle fibres converge and appear to attach to both the outer and the inner layers. The precise ultrastructural organization of non-localized centric systems appears to be somewhat variable. In *Rhodnius,* a diffuse centromere is spread along the whole leading edge of the mitotic chromosome. In *Luzula, Cyperus,* and *Dysdercus,* each chromosome appears to be genuinely polycentric with a series of localized spindle-binding sites. In *Oncopeltus,* centromere plates occupy up to 75 per cent of the long axis of the chromosome and have the same tripartite structure as that seen in localized centric systems. In all these cases, however, the meiotic chromosome is quite different in its organization. There are no structures even vaguely resembling centromere plates, and chromosome movement appears to be mediated by the microtubules simply inserting into all parts of the chromosomes.

(b) *The nucleolar organizer.* In addition to the primary or centric constriction, chromosomes with localized centromeres may also have secondary regions which do not coil in company with the main body of the chromosome (Figs. 7 and 14). The only known function of such secondary constrictions is nucleolar organization. At one time the nucleolus was thought to be a nuclear receptacle which accumulated RNA synthesized elsewhere. This proposed function led to the term 'nucleolar organizer' being given to the locus at which this organelle formed. The more recent evidence is that in animal cells, at least, the nucleolus is the site of a 45S rRNA precursor. Thus, the organizer represents a specific sequence of DNA, which codes for rRNA, that can be referred to as rDNA. In eucaryotes three kinds of rRNA are in fact known; they have different sedimentation coefficients, namely, 5S, 18S, and 28S. The 5S class does not appear to be coded in the rDNA of the organizer region in either *Drosophila melanogaster* or *Xenopus laevis.* In both these species, however, the organizer consists of tandemly duplicated copies of a pair of genes which transcribe for the 45S RNA which is later split into 18S and 28S molecules. These represent the RNA components of the smaller and larger ribosome subunits respectively (see Table 3). In *Drosophila melanogaster* there are about 130 copies of these cistrons while in *Xenopus laevis* there are some 460–650 copies (Table 4). Although the genes for 5S RNA are not closely linked to the 28S and 18S genes, the synthesis of all three is co-ordinately controlled. In *Xenopus* there are at least 9000 copies of the 5S gene. By *in situ* hybridization, Pardue, Brown, and Birnstiel (1973) have demonstrated that the 5S genes are present at the end of the longer arm of most if not all the chromosomes.

Nucleolar regions can be fractured experimentally by X-irradiation.

The subunits so produced remain functional, which supports the view of the multiple nature of the organizer region. It thus includes both complementary and supplementary components.

(c) *Heterochromatic differentiation.* Primary and secondary constrictions are discernible only in relatively condensed chromosomes (late prophase to anaphase). During early prophase, however, a third distinctive region can be distinguished, which stands out by virtue of its advanced state of condensation. These heterochromatic regions appear also in the preceding interphase as heteropycnotic blocks which are often aggregated into so-called chromocentres (Figs 5 and 14). Heterochromatin is thus that chromatin which remains compact during interphase of the cell cycle when the remainder, the so-called euchromatin, is more dispersed. The two types of chromatin also stain differentially. The molecular mechanisms which underlie this heterochromatinization in fact control the degree of condensation or coiling of the chromosome and so render the DNA strands inaccessible for RNA transcription (see p. 88). Heterochromacy frequently characterizes regions adjacent to centromeres, nucleolar organizers, or chromosome ends (telomeres), while two classes of chromosomes often show this character in their entirety, namely sex and supernumerary chromosomes.

Brown (1966) has subdivided heterochromatin into two categories:
(*i*) constitutive—which reveals its heterochromatic nature at virtually all stages in almost all cell types; and
(*ii*) facultative—which may vary in state in different cell types, or different developmental stages or even between homologues in the same cell.

In the case of facultative heterochromatin, positive heteropycnosis reflects a functionally inert state temporarily assumed by a particular chromosome. The best example of this is the mammalian sex or X-chromosome. In normal females one of the two X-chromosomes is regularly heterochromatic in somatic cells (see p. 73). In female germ cells, on the other hand, both X-chromosomes are euchromatic. When either of the X-chromosomes is transmitted to male progeny it remains euchromatic. Thus this chromosome can assume a heterochromatic or a euchromatic state in response to the particular environment in which it finds itself.

In constitutive heterochromatin, on the other hand, the tendency for positive heteropycnosis is claimed to be inherent in the particular base sequence contained in the DNA of the region in question. The regions adjacent to the centromeres of mouse chromosomes maintain their heterochromatic character at all times and are regarded as constitutive. Hybridization of extracted radioactive material with the DNA of cytological preparations shows that in mouse, sequences of satellite DNA are located in this centric heterochromatin. Gall, Cohen, and

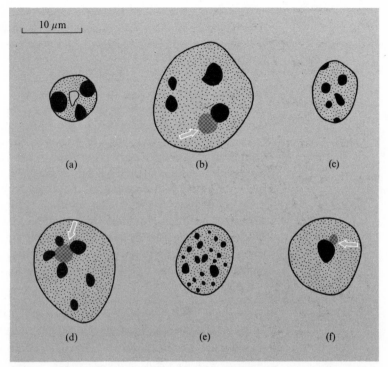

FIG. 5. Heterochromatin variation in mouse cells (after Hsu, Cooper, Mace, and Brinkley (1971)).
- (a) Granulocyte from bone marrow, small number of large heterochromatic blocks.
- (b) Neuron from cerebellum, small number of large blocks plus a prominent nucleolus (arrow).
- (c) Glial cell from cerebellum, numerous small blocks.
- (d) Sertoli cell from testis, a few large heterochromatin blocks associated with a large nucleolus (arrow).
- (e) Spermatogonial cell, large number of very small blocks.
- (f) Spermatid nucleus, single large block of heterochromatin associated with a nucleolus (arrow)

Polan (1971) have also localized satellite DNA in the centric regions of *Rhynchosciara, Triturus,* and *Drosophila.* This association of highly repeated sequences with constitutive heterochromatin may well be important, since genetic material located in heterochromatin is not usually expressed (see p. 73). Mouse chromosomes have also been held to contain constitutive heterochromatin at their ends, or telomeres, as well. If these regions also contain satellite sequences then they are not sufficiently numerous to be detected by the cytological hybridization technique. The Y-chromosome of the mouse, which also contains procentric heterochromatin, is devoid of satellite segments. Clearly, heterochromatin is not a single category with respect to satellite content either.

This is particularly well demonstrated in *Drosophila melanogaster*. In salivary glands of this species, two morphologically distinct fractions of heterochromatin are represented. One type, the α-heterochromatin, forms a small compact mass at the mid point of the chromocentre. The second type, the β-heterochromatin, forms a larger, loosely textured, and poorly banded region, which extends into the chromosome arms. The α-heterochromatin corresponds to the bulk of the mitotic hetero-chromatin and is not replicated during the polytenization process which occurs in the gland (see p. 35). The β-heterochromatin, on the other hand, does replicate during the formation of the polytene nucleus but represents only a relatively small part of the mitotic chromosome. Different species of *Drosophila* differ both in the size of the centromeric heterochromatin block, as seen in mitotic chromosomes, and in the amount of satellite DNA. Moreover, in terms of the available data, there appears to be a correlation between the amount of satellite DNA present in a species and the amount of centromeric heterochromatin in the mitotic chromosomes of that species.

2.1.3. DNA variation between species

The amount and precise informational content of DNA in any species must relate to its requirements in development, metabolism, and re-production. As organisms increase in complexity, all three requirements are expected to increase and, on a broad view, the amount of DNA also increases with increasing complexity (Fig. 6, Table 6). Even so, there are very clear cases where the amount of DNA per nucleus does not provide a simple index to either structural or functional complexity. Thus most of the biosynthetic pathways which are known to be present in complex organisms are found in simple ones too and over 90 per cent of the known enzymes appear to be common to both pro- and eucaryotes.

Of course, as we have already seen (p. 11), there is a measure of repetitiveness in the DNA of eucaryotes, and one very obvious function of such repetition might be to provide for an increase in the number of regulatory units required to control the developmental process. This is not likely to be the case for the highly repetitive fractions but could well apply to the so-called intermediate fractions. In these terms it can be argued that the principal difference between the simpler and the more complex eucaryotes lies not so much in the number of structural genes they contain as in the increased complexity of gene regulation required to support cell differentiation and morphogenesis (see p. 62). Although it is commonly claimed that repeat sequences are not transcribed, a number of studies indicate that RNA sequences are formed from repetitive DNA both during development and in differentiated adult cells. Many of these RNA sequences, however, do not leave the nucleus, so that their function is not known. Certainly it seems unlikely that some satellite sequences code for an RNA capable of specifying the

Table 6

DNA content per haploid genome of some selected eucaryotes

Species	Haploid chromosome number	Nucleotides per haploid cell
1. *Neurospora crassa* (ascomycete fungus)	7	8.6×10^7
2. *Chlamydamonas reinhardi* (green alga)	15	1.2×10^8
3. *Drosophila melanogaster* (fruit fly)	4	1.7×10^8
4. *Ciona intestinalis* (sea squirt)	14	3.4×10^8
5. *Ustilago maydis* (basidiomycete fungus)	2	3.8×10^8
6. *Chironomus tentans* (dipteran fly)	4	5×10^8
7. *Marchantia polymorpha* (bryophyte)	9	9×10^8
8. *Amphioxus lanceolatus* (cephalochordate)	12	1×10^9
9. *Esox lucius* (pike)	9	1.7×10^9
10. *Strongylocentrus purpuratus* (sea urchin)	18	1.8×10^9
11. *Columba livia* (pigeon)	40	2×10^9
12. *Macropus rufus* (kangaroo)	10	3.1×10^9
13. *Boa constrictor* (snake)	18	3.5×10^9
14. *Epatretus stoutii* (hagfish)	24	4.7×10^9
15. *Alligator missipiensis* (alligator)	16	5×10^9
16. *Selaginella kraussiana* (bryophyte)	10	5.1×10^9
17. *Euglena gracilis* (green alga)	45	5.8×10^9
18. *Mus musculus* (mouse)	20	6.5×10^9
19. *Gryllus domesticus* (cricket)	11	1.1×10^{10}
20. *Vicia faba* (broad bean)	6	4.4×10^{10}
21. *Pteridium aquilinum* (fern)	52	4.4×10^{10}
22. *Triturus cristatus* (newt)	12	4.5×10^{10}
23. *Pinus banksiana* (gymnosperm)	12	7.5×10^{10}
24. *Protopterus* (African lungfish)	17	1.0×10^{11}
25. *Lepidosiren paradoxa* (lungfish)	19	1.1×10^{11}
26. *Necturus maculosus* (mudpuppy)	12	1.7×10^{11}
27. *Lilium longiflorum* (lily)	12	1.8×10^{11}
28. *Amphiuma means* (congo eel)	12	1.9×10^{11}

Note. Data from Sparrow, Price, and Underbrink (1972).

primary structure of a protein, but this does not exclude the possibility that the RNA produced is regulatory in function.

There are, however, two major complications to this kind of argument.

(*i*) The proportion of the genome that is repetitious may vary considerably even between closely related species. This applies particularly to the satellite fraction where large differences are found both in quantity and composition. For example, the three closely related species *Drosophila hydei, D. neohydei,* and *D. pseudohydei* all have satellites of different bouyant densities and of different magnitudes. This suggests that satellite sequences arise frequently in evolution and that whatever function or functions they have can be undertaken by many, though not necessarily all, kinds of sequence.

(*ii*) Some organisms have disproportionately large amounts of DNA

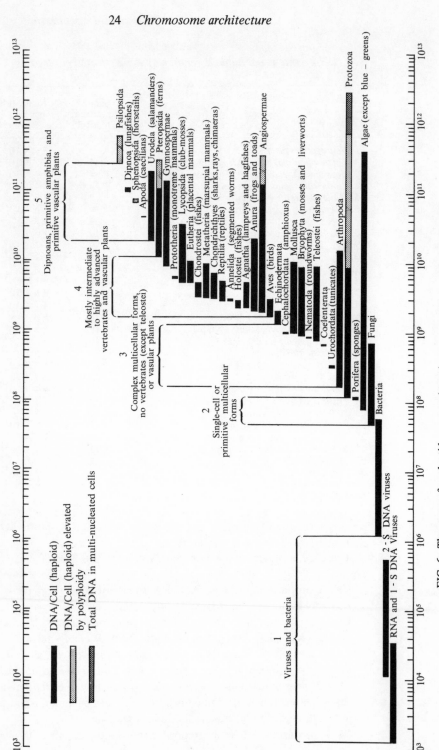

FIG. 6. The range of nucleotide content in major taxonomic categories arranged according to increasing minimum amount (data from Sparrow, Price, and Underbrink (1972)).

which cannot be accommodated in terms of highly repetitive DNA.
This applies particularly to lungfish (Dipnoi) and urodele amphibians,
among animals (Fig. 7). In addition there is an inordinately wide
range of DNA values within certain phyla, even among species within
a single genus, which are therefore of comparable complexity.
This implies that large-scale changes in DNA do not necessarily reflect
substantial genetic divergence.

Three mechanisms are known which are capable of increasing the
quantity of DNA in an organism—duplication of genes or gene segments,
additions of whole chromosomes (aneuploidy), and the multiplication
of whole chromosome sets (polyploidy).

Aneuploids are rare in nature since the imbalance consequent upon
the 'asymmetrical' gain or loss of whole chromosomes is almost always

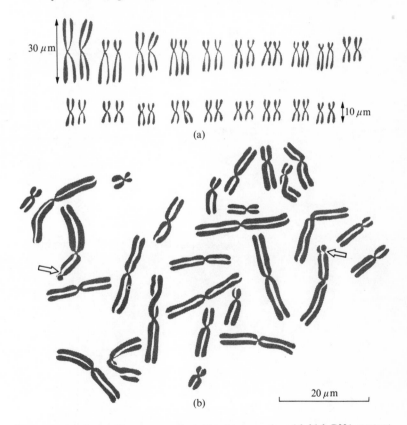

FIG. 7. Mitotic chromosome complements of two species with high DNA content.
(a) South American lungfish, *Lepidosiren paradoxa*, female ($1 \cdot 1 \times 10^{11}$
    nucleotides per haploid cell) (after Ohno and Atkin (1966)).
(b) The axolotl or tiger salamander, *Ambystoma tigrinum* ($8 \cdot 3 \times 10^{10}$
    nucleotides per haploid cell). Nucleolar organizers marked by arrows
    (after Callan (1972)).

developmentally deleterious. Supernumerary or B-chromosomes constitute an exception to this statement but significantly these are usually
much modified and are often heterochromatic in character. By contrast
polyploidy is widespread, especially in plants. As far as the quantity of
DNA in individual chromosomes is concerned it can, in theory, be
increased in two rather different ways (Fig. 8). First, by a lateral
multiplication in the number of DNA dimers per chromosome (differential polynemy); and secondly, by lengthwise repetition of chromosome
segments (tandem duplication). Differential polynemy demands that
the chromosomes of at least some species are multistranded, or polynemic,
embodying two, or multiples of two, dimers running parallel along the
length of each chromatid. Most contemporary workers believe that the

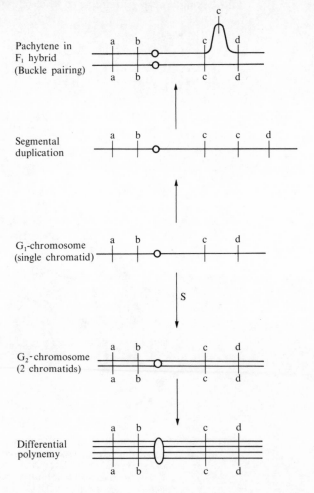

FIG. 8. Mechanisms for producing intrachromosomal variation in DNA.

chromatid is essentially a single DNA dimer and do not accept that chromosomes are or can be polynemic. Tandem duplication is therefore the conventional mechanism assumed to account for DNA increases within chromosomes.

In brain ganglion cells of *Drosophila melanogaster,* the diploid chromosome number is maintained throughout development but the chromosomes are uniformly longer and thicker in third-instar larvae than at first instar. The chromosomes, however, are morphologically identical at both stages and there is no alteration in the hetero-chromatin—euchromatin ratio. They are not diplochromosomes but there is a twofold difference in DNA content and the cells are also proportionately larger. Gay, Das, Forward, and Kaufmann (1970) have concluded that the cells from the later developmental stages are probably polytene; and if polyteny is possible in development it may also occur in evolution.

Given that polyploidy and tandem duplication appear to be the main mechanisms for increasing the quantity of DNA in the nucleus, let us examine how much divergence in DNA value they are capable of achieving.

In the plant genus *Allium*, there are both diploid and polyploid species, and the former include types with 14, 16, or 18 chromosomes in the diplophase. By estimating the relative DNA content of a range of these species using Feulgen photometry, Jones and Rees (1968) have shown that:

(*i*) a constancy of chromosome number does not reflect a constancy of nuclear DNA;

(*ii*) basic chromosome number is not a reliable guide to DNA content. Thus some of the 14-chromosome diploids have higher DNA values than the 16- or 18-chromosome diploids and some of the 18-chromosome diploids have more DNA than some of the 32-chromosome tetraploids (Fig. 9).

(*iii*) The range of DNA content among the diploids may be as great as that resulting from polyploidy.

By producing an F_1 hybrid between *A. cepa* and *A. fistulosum*, they were able to show that duplication-deficiency differences do indeed exist between these two diploids as evidenced by buckle pairing at pachytene and the production of asymmetrical bivalents at first metaphase of meiosis. Over-all, *A. cepa* has about 27 per cent more DNA than *A. fistulosum* and in the F_1, all the first metaphase bivalents are visibly asymmetrical, the degree of asymmetry varying from 60 per cent to 20 per cent. These results suggest that the differences which do exist between the diploid forms are, indeed, the result of duplication. If this is the case then the elevation of DNA content that can acrue from duplication at the diploid level can be very considerable indeed.

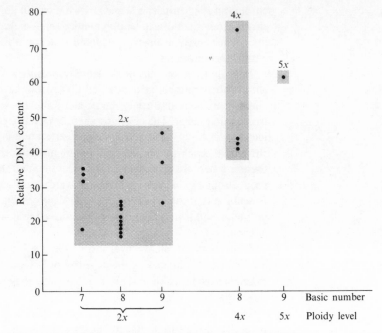

FIG. 9.   DNA variation in *Allium* (after Jones and Rees (1968)).

Thus DNA values alone cannot distinguish adequately between duplication and polyploidization.

It will be appreciated that in the case of related diploid-polyploid
comparisons, the direction of evolutionary change can be predicted
with some confidence. Diploids give rise to polyploids and can rarely
be expected to arise from them. However, where DNA differences at
the same ploidy level are concerned, evolutionary increases (by duplication) or decreases (by deletion) may be involved. In such cases only
rarely is there a basis for showing preference for a particular direction
of evolutionary change. For example, a cytochemical study of hybrids
between the flies *Chironomus thummi thummi* and *C. thummi piger*
has shown that certain bands in the polytene chromosomes (see p. 35)
of *C. thummi thummi* differ in DNA content from those of *C. thummi
piger* by factors of 2, 4, 8, or 16. That is, an eightfold difference in
DNA can occur between corresponding bands of related sub-species, and
the pattern of elevation implies a progressive but localized doubling of
DNA content. Over 20 bands show this relationship. They occur on
both sides of the centromeres of the three larger pairs of chromosomes
and contribute to an over-all difference in DNA content of some 27 per
cent. In many cases different geometrical ratios are found for the same
region in different individuals.

The fact that many loci in *Drosophila* appear on both cytological and genetical grounds to be the consequence of duplication, triplication, or even higher-order multiplication, lends support to the possibility that a considerable increase in DNA content could result from the duplication of small segments alone. Where, as in vertebrates, the requisite techniques for the analysis of genetic fine structure are not available, the presence of undiscovered linear repetition of cistrons specifying a variety of proteins cannot be excluded. Lungfish and urodele amphibians, for example, both show exceptionally high DNA values and exceptionally large chromosomes. Both, however, have a relatively modest chromosome number with no suggestion of a polyploid origin (Fig. 7). In the urodeles, genomes with a high DNA content also have a high proportion of repeated sequence DNA (Table 7) so that, on balance, the increase seems to stem from duplication.

Table 7
Repeated sequence DNA in amphibians

Organism	pg DNA per haploid genome	Nucleotide pairs	% of genome which is repeated
Scaphiopus	1·2	$1\cdot1 \times 10^9$	40
Engystomops	2·7	$2\cdot5 \times 10^9$	50
Xenopus	3·0	$2\cdot7 \times 10^9$	45
Bufo	5·0	$4\cdot6 \times 10^9$	80
Rana	8·0	$7\cdot3 \times 10^9$	78
Ambystoma	42·0	$3\cdot8 \times 10^{10}$	80

Note. After Britten and Davidson (1971).

The most obvious process which could lead to duplication is faulty replication. That this can indeed lead to substantial differences in DNA in a short space of time is clear from studies involving the plant genus *Nicotiana. N. otophora* is a diploid species which has large hetero-chromatic blocks in 5 of the 12 chromosome pairs of the complement. The tetraploid *N. tabacum*, on the other hand, has only scattered hetero-chromatin. When an *otophora* chromosome with a large heterochromatic block is introduced into the *tabacum* complement by hybridization it behaves abnormally and in a small proportion of the cells gives rise to chromosomes of astonishing length, which are largely heterochromatic. Such unusually large chromosomes (megachromosomes) appear to arise from extra replication of the heterochromatin material they contain.

Yet a further example of the same principle has recently been described in *Drosophila melanogaster*. Here flies that contain fewer rRNA genes, i.e. less rDNA, than the wild-type exhibit a mutant phenotype known as bobbed (bb). Bobbed flies have smaller bristles, a thinner cuticle, and a reduced growth rate. Thus bobbed is essentially a partial deficiency for rDNA and mutants of this type are known in both the X- and Y-chromosomes since each contains a single nucleolus organizer.

In wild-type females there are about 250 rRNA genes in the nucleolar organizer of each X-chromosome. When such an organizer is present in flies in only a single dose, as in either X/O males or else in experimentally constructed females deficient for the nucleolar organizer (NO) in one of the X-chromosomes (X/XNO⁻) then the number of rRNA genes in the normal X increases to about 400. In addition rRNA genes of bobbed mutants also undergo disproportionate replication when maintained with a homologue that is completely (XNO⁻) or partially (Ybb⁻) deficient in its rDNA. Indeed the extent of the deficiency determines the extent of disproportionate rDNA replication in the bobbed mutant organizer.

Such disproportionate replication can apparently occur in the soma only (bb⁺/O♂, bb⁺/XNO⁻♀, bb⁺/Ybb⁻♂) or else in both soma and germ line (bbᵐ/XNO⁻♀ and bbᵐ/Ybb⁻♂) but only in the case of bbᵐ/Ybb⁻ males are these increased rRNA genes structurally integrated into the genome and inherited in a stable manner.

There are potential difficulties in incorporating multiple copies of a gene segment into the same genome. In the first instance it affects the over-all genetic balance of the nucleus. It also invites unequal crossing-over and so it could lead to instability of the genome (Fig. 10). Polyploidy has neither of these inherent disadvantages of tandem duplication though it does introduce other problems. Further, where the several copies of a region are contained on separate chromosomes, functional diversification of the multiple loci can be generated more easily.

Ohno (1970), in particular, has stressed that polyploidization is the most effective means of making new genetic material available for

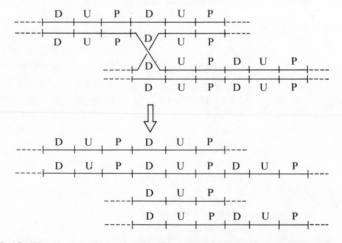

FIG. 10. The consequences of unequal crossing-over in a tandem duplication.

evolution. His argument is based largely on a consideration of DNA relationships in vertebrate animals. From such considerations he has suggested that polyploidy has been involved in vertebrate phylogeny on at least four occasions.

(a) In *lampreys*, tetraploidization of a presumed ancestral diploid genome containing 48 rod-chromosomes has been assumed to have led to the development of forms with increased DNA content. But quite independently, hagfish have increased their DNA to a much higher level without any change in diploid number, presumably by extensive duplication.

(b) In *teleost fish* two presumptive cases of tetraploidization have been claimed.

(*i*) In the order Isospondyli there are two closely related suborders, the Cupeiodea and the Salmonoidea. The former and the smelts among the latter are believed to be diploid ($2n = 48-52$), but the Salmonidae has been claimed to be tetraploid because, though the chromosome number ranges only from $2n = 56-64$, there is approximately a tetraploid number of chromosome arms (104 in presumed tetraploids compared with 48–60 in diploids), each chromosome being two-armed. There is also twice the amount of DNA per nucleus ($6 \cdot 0 \times 10^{-9}$ v. $3 \cdot 0 \times 10^{-9}$ mg maximum).

In line with this, salmonoids have more representatives of some gene loci than any other group of vertebrates. For example, salmon and trout have two loci each for trypsin and chymotrypsin, which is markedly unusual in vertebrates. Again cupeoids and smelts have two or three loci for subunits of LDH whereas trout, salmon, and grayling have at least five and possibly as many as eight separate loci. Finally the presence of meiotic multivalents ranging from associations of III to associations of VI also supports the occurrence of tetraploidy in this group (see p.150).That multivalents larger than IV can form is explained by the fact that intra-individual polymorphism for Robertsonian fusions (see p.133)has been demonstrated in the group, so that several different diploid cell populations varying in chromosome number can occur in a given individual.

(*ii*) Within the family Cyprinidae, the carp (*Cyprinus carpio*) and the goldfish (*Carassium auratus*) stand out by having a chromosome number of 104 while other members of this family show $2n = 44-54$. Ohno (1970) again concludes that both these species are tetraploid since their DNA values are also approximately twice those of other members of the family.

Supporting evidence is provided by the fact that these presumptive tetraploids have more gene loci than their diploid counterparts. During meiosis, however, only bivalents are formed, so that a secondary diploidization has been superimposed on the tetraploid system (see

p. 151) whereas in the Salmonidae the process of diploidization is not yet complete.

(c) Polyploidy appears also to have arisen in *anuran amphibians*. Indeed, despite objections which have been raised against the possibility of bisexual polyploidy, seven polyploid and biparental anuran species are now known (Table 8). For example, in the family Ceratophyridae the genus *Odontophrynus* includes both diploid species like *O. cultripes* and *O. occidentalis* with $2n = 2x = 22$ and tetraploid forms like *O. americanus* with $2n = 4x = 44$. Here the chromosomes fall into 11 groups of homologues, each group having 4 members. At meiosis, up to 11 quadrivalents may form, and both DNA content and nuclear volume comparisons support the suggestion of simple tetraploidy. Interestingly, however, not all members of *O. americanus* are tetraploid. Diploid forms also occur which are indistinguishable both in external characters and in size although they have only half the amount of nuclear DNA. In a second genus, *Ceratophrys,* the diploid *C. calcarata* has $2n = 26$ but *C. dorsata* has $2n = 104$ and the chromosomes can be arranged into 13 groups of 8 homologues each. Primary spermatocytes in *dorsata* contain a mixture of octavalents, hexavalents, quadrivalents, and bivalents. Multivalents occur also in *Phyllomedusa burmeisti* and *H. versicolor* but not in either of the *Pleurodema* species. *H. versicolor* and *H. chrysocelis* are distinguished only on the basis of bio-accoustic analysis, slow-trilling animals are $4x$, fast-trilling animals are $2x$. A pulse-rate difference in call holds also in *O. americanus* where, again, the $4x$ form has a reduced pulse rate. Speciation by polyploidization thus appears to be an important evolutionary mechanism in the Anura.

How then can polyploidy arise in bisexual sytems? One possibility is that it orginates indirectly by polyploid parthenogenesis followed by a secondary reversal to sexuality. This can be thought of as a four-step process:

(*i*) the induction of diploid parthenogenesis;

(*ii*) the adoption of tetraploid parthenogenesis by polyploidization of the diploid parthenogenetic form;

(Actually these two steps might well occur simultaneously since in several cases where parthenogenetic races have developed polyploidization has taken place at the moment of transition from amphimixis to parthenogenesis.)

(*iii*) the formation of a triploid parthenogenetic form by the crossing of the tetraploid female parthenogenetic form with a diploid male; and, finally

(*iv*) a reversal of the triploid to a bisexual mode of reproduction with the preservation of polyploidy. Thus if the parthenogenetic triploid where to produce occasional hexaploid ($6x$) primary oocytes which underwent a normal meiosis, this would give rise to a reduced

Table 8

Polyploidy in anuran amphibians

Species	2n	DNA content			Ploidy
		Relative value	Absolute value per haploid genome	Percentage mammalian level content	
(1) Ceratophrididae					
Odontophrynus carralhoi	22	50	5×10^9	45	2x
occidentalis	22	50	5×10^9	45	2x
cultripes	22	65	$6 \cdot 6 \times 10^9$	59	2x
americanus	44	107	$1 \cdot 1 \times 10^{10}$	97	4x
F₁ *O. cultripes* × *O. americanus*	33	72		65	3x
Ceratophrys calcarata	26	59	6×10^9	54	2x
dorsata	104	299	2×10^{10}	181	8x
(2) Hylidae					
Hyla parkeri	24	74	$7 \cdot 4 \times 10^9$	67	2x
faber	24	86	$8 \cdot 7 \times 10^9$	78	2x
multilineata	24	117	$1 \cdot 2 \times 10^{10}$	106	2x
pulchella prasina	24	162	$1 \cdot 6 \times 10^{10}$	147	2x
chrysoscelis	24				2x
versicolor	48				4x
F₁ *H. chrysolscelis* × *H. versicolor*	36				3x
Phyllomedusa burmeistei	52	162	$1 \cdot 6 \times 10^{10}$	147	4x
Standard: *Triturus cristatus*	24	448	$4 \cdot 5 \times 10^{10}$	407	2x

Note. Data of Beçak, Beçak, Schreiber, Laralle, and Amonum (1970); note absolute DNA values have been estimated from cytophotometric data using the urodele *T. cristatus* as a reference point.

34 *Chromosome architecture*

triploid egg which after fertilization with haploid sperm from a
diploid male would give rise to tetraploid zygotes of both sexes.
Significantly as we have already seen diploid forms still exist in
O. americanus.

(d) Finally Ohno, Wolf, and Atkin believe that the phylogenetic line
which gave rise to the amniotes was also tetraploid, so that reptiles,
birds, and mammals are all basically tetraploid lineages which have become secondarily diplodized so far as their meiotic behaviour is concerned. Thus surviving members of the Reptilia have about the same
amount of DNA as mammals, and so do birds.

The over-all picture that emerges from vertebrate studies is that, while
polyploidization has certainly occurred, the species with the highest
DNA content are not of polyploid origin. As far as urodele amphibians
are concerned, a higher proportion of repeat sequence DNA is present
in genomes of larger size (Table 7). Even so, there remain some puzzling
problems. Thus although only 20 per cent of the genome of *Amphiuma*
contains unique sequences, this still means that *Amphiuma* must have at
least twelve times the number of unique sequences as man, where 50 per

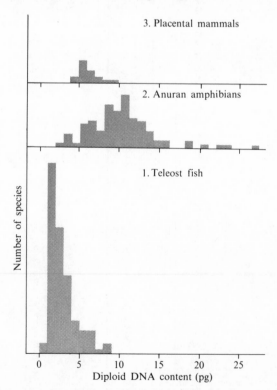

FIG. 11. Diploid nuclear DNA values in three vertebrate groups (after Bachmann,
Goin, and Goin (1972)).

cent of the DNA is unique. Indeed the restricted range of mammalian DNA values (Fig. 11) is particularly puzzling in view of the wide qualitative difference found in mammalian species.

2.1.4. *Chromosome activity*

(a) *Transcriptive functions.* In dipteran flies, initial embryonic growth is regular but the cell number in larval tissues is fixed. No mitoses occur after organ formation, rather the cells then grow by an endomitotic process during which the chromosomes replicate 8, 9, or even 10 times. As the cell grows, the chromosome number does not increase although the DNA undergoes an approximately geometrical increase. Instead the products of each successive replication cycle remain together. This, coupled with an initimate somatic pairing of homologous chromosomes, which characterizes most dipterans, results in the many homologous chromatid equivalents forming an essentially single polytene chromosome. An identical process occurs also in collembolans.

Two features which accompany polyteny are increased length and width on the one hand and the development of a system of transverse banding on the other. The latter involves the formation of a series of chromatic zones, or bands, separated by achromatic interband regions. The bands differ from one another in appearance and their specific sequence can be mapped consistently throughout the greater part of the complement (Fig. 12). By comparing linkage maps with banding maps, a correspondence can be established between particular gene functions and specific bands. Thus the mutant 'notch', which behaves genetically as a small deficiency, involves the loss of band 3C7 from one homologue only in heterozygous notch females. In no case has it been necessary to allocate more than one Mendelian gene to any one band and there has been a tendency to assume an absolute equivalence of bands and genes.

Polytene chromosomes are developed in those organs which are engaged in vigorous metabolic activity and they have been found in salivary glands, malphigian tubules, mid- and hind-gut epithelia, ovarian nurse cells, bristle initial cells, and foot-pad cells. In all of these cell types the bands can exist in one of two states. First, compact and condensed and secondly, diffuse and swollen. A swollen band is referred to as a puff, or in extreme cases as a Balbianni ring (Fig. 12). And all the available evidence indicates that puffing is a reflection of increased gene activity.

The usual basis for deciding whether a particular gene is active or not is to determine whether the final product of that gene is present or absent. Bands contain high concentrations of both DNA and histone, while interbands have relatively low concentrations of these chromosome components. In general, the amount of DNA and histone appears to remain unchanged during the transformations which convert the compact

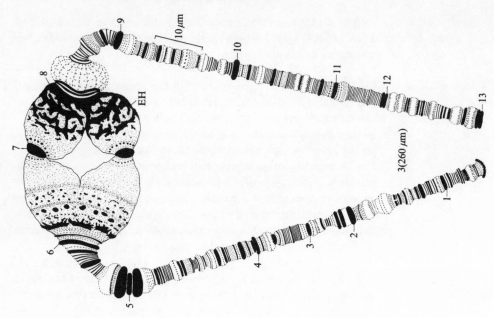

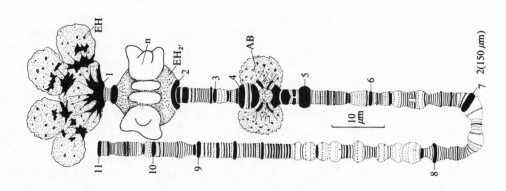

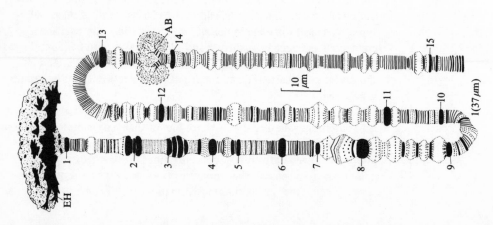

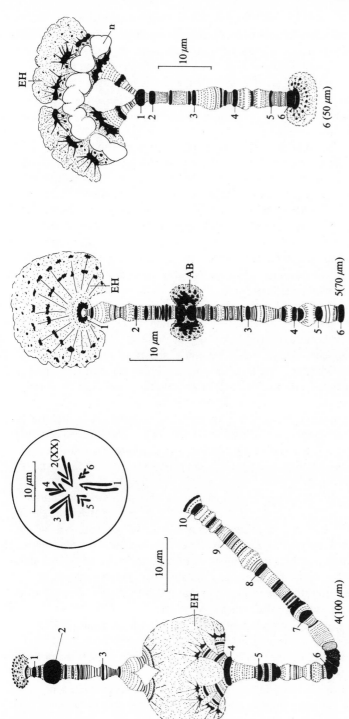

FIG. 12. The polytene salivary gland chromosomes of the collembolan *Bilobella grassei* ($n = 6$) from the posterior part of the gland. Even within the same population these chromosomes show considerable variability in the expression of the bands, the heterochromatic regions (EH) and the Balbianni rings (AB). Note that not all the chromosomes are drawn to the same scale. The position of the centromere is not known with certainty in the polytene elements; n = nucleolar ring. For comparison a mitotic complement is also shown (inset circle). (after Cassagnau (1971)).

band into a diffuse puff. Puffs, however, also contain significant amounts of RNA and the balance of evidence indicates that this is mRNA. Thus, in the first place, its AU and GC ratios are unequal, suggesting that this RNA was templated by a single DNA strand. In the second place, ribonucleoprotein particles can be shown to be present in electron-micrographs of the puff. They can also be found free in the nuclear sap and can be seen passing through the pores of the nuclear membrane into the cytoplasm.

That the situation really is as simple as this has been elegantly demonstrated in *Camptochironomus*. Here a distinctive group of four cells is found near the duct of the salivary gland. In *Ch. pallidivattatus* these four cells produce a granular secretion, the so-called SZ granules, while in *Ch. tentans* the same cells give a clear non-granular fluid. This difference is associated with a difference in puffing activity in chromosome IV, a particular puff being present in *pallidivittatus* but not in *Ch. tentans.* In hybrids between the two species the puff appears only in the homologue derived from *pallidivittatus,* so that a heterozygous or half-puff is formed. Significantly this half-puff produces a far smaller number of SZ granules. Heterozygous puffs are also known in certain mutant conditions, the band from the mutant stock remaining inactive in wild type/mutant (+/m) heterozygotes.

In condensed bands the DNP material assumes a complex tertiary structure whereas in the interband regions it may be aligned more or less in parallel with the long axis of the polytene chromosome. During puff formation the complex tertiary structure is simplified by a separation and extension of the material of the band. This extension is a necessary prerequisite for transcription. Thus a gene segment can be regarded as inactive when its position on the polytene chromosome is defined by a condensed band, but in an active state when that band is puffed.

Now although bands appear to be equatable with single structural cistrons, it has been estimated that an average band contains something like 10^{-16} g DNA. This is equivalent to some 100 000 base pairs which represents enough information to code for about 30 000 amino acids. This is improbably high for the product of a single cistron and implies that, either bands represent operational units of greater complexity than has hitherto been assumed, or else that a large part of this DNA is repeated or transcriptively silent.

If the genetic organization of DNA within the bands of polytene chromosomes is complex and if there is a linear repetition in informational content, then puffs might be expected to behave in a manner similar to nucleolar organizers in the sense that a break within a band would give rise to two elements each capable of organizing a separate puff. In two cases this has been shown to be so. In the chironomid

Smittia, a puff in the salivary gland nucleus has been divided successfully by irradiation into two separate puffs. Further, Ashburner (1970) has shown that an X-ray-induced translocation, SC^{260-15}, in *D. melanogaster,* breaks the thread in the region of a very large puff, 71CE, and in translocation heterozygotes two puffs are found whose activity is co-ordinate during development.

(b) *Replication mechanisms.* The semiconservative nature of DNA replication in procaryotes is well established and appears to apply also to the DNA of chloroplasts and mitochondria in eucaryotes where the molecule, as in many procaryote systems, is a simple circular dimer. It has generally been assumed that the chromosomes of eucaryotes also replicate semiconservatively although the evidence in this case is less definitive. How semiconservative replication is possible within complex chromosomes containing relatively enormous amounts of DNA is something of an enigma.

Unique features of procaryotic chromosomes are their singularity, their circularity, and their lack of histones. Such simplicity is mirrored in their simple mode of replication, since most of the evidence to date suggests that the whole genome constitutes a single unit of replication or replicon. In other words, replication is initiated at a single site in the genome from which it proceeds until the whole unit has been reproduced. Some evidence indicates unidirectional synthesis in which case there would be only one replication fork which moves around the circular chromosome doubling the DNA dimer as it goes. This fork starts at an origin and finishes at an adjacent terminus. In most strains of *E. coli* there is a fixed origin lying in the lower left quadrant of the conventional chromosome map with replication proceeding clockwise. Variant strains are known, however, where the origin is located elsewhere and where replication may even be anti-clockwise. Recently Masters and Broda (1971) have suggested that replication in *E. coli* may in fact proceed sequentially in both directions from the origin. If this proves to be true then it brings the procaryotic pattern more into line with that found in eucaryotes, though the procaryotic chromosome could still be regarded as a single replicon.

A further feature of the replication process in procaryotic cells is that the pattern of replication can vary according to the growth rate of the cell in question. In fast-growing cultures it is continuous and in this respect contrasts markedly with the periodic nature of DNA replication in eucaryotes. Here replication is confined to a restricted period of about 6–8 hours which occupies about a third of the total interphase period. Thus bulk DNA synthesis in eucaryotes is completed well ahead of the division cycle whereas in procaryotes the two processes can be coincident.

Through the use of tritium-labelled thymidine ($[^3H]$ TdR), a specific precursor in DNA synthesis, it has been shown that not all parts of the chromosome label synchronously during DNA synthesis (S period). Rather there are multiple initiation sites of synthesis along each chromosome so that there is a tandem or serial arrangement of many relatively short replicons along the chromosome. Not only do eucaryote chromosomes contain many replicons but these appear to be arranged differently in different organisms, for there is no consistent order of synthesis in chromosomes of different species.

Because the eucaryotic chromosome has multiple initiation sites for DNA synthesis, there has been a tendency to assume that the individual units of replication are separated from one another by non-DNA linkers so that there must be many free ends of DNA. In fact there is no convincing evidence for either assumption. Indeed in procaryotes, where the DNA is arranged in a closed circle, the initiation of DNA synthesis, although occurring at a specific site, does not require permanent free ends or attachment of non-DNA linker material. Likewise, as we shall see shortly (p. 65), the DNA present in free, floating copies of nucleolar organizing regions produced by selective replication are also in the form of circles of varying size which may well consist of a variable number of component regions.

While the pattern of autoradiographic labelling shows that there are multiple replication points along the eucaryote chromosome, they do not give much accurate information on the number of points or the mode of replication. These questions have, however, received some attention through the autoradiographic study of individual DNA molecules. Such studies indicate that each replicating section has two growing points which move in opposite directions from a common origin (Fig. 13). The commonest length of a single replicon is of the order of 30 nm, but many are shorter. These results thus suggest a bidirectional model for DNA replication. Neighbouring sections can begin replication at different times and, at any one time, the regions of synthesis are not distributed equidistantly along the fibres. Thus replication is also asynchronous.

The most pronounced form of asynchrony relates to heterochromatic chromosomes or chromosome regions which are regularly late-labelling and which replicate faster than euchromatic regions. One of the best-known cases of such label asynchrony is that found in one of the two X-chromosomes which are present in the majority of female placental mammals. In this group there is, as a rule, an XY♂ XX♀ sex-determining system. Mammalian X-chromosomes vary in shape, and their form, in some measure, is related to the fraction of the haploid complement which they represent. The simplest and most common of these, and the one conventionally considered as standard, is that where the X/autosome

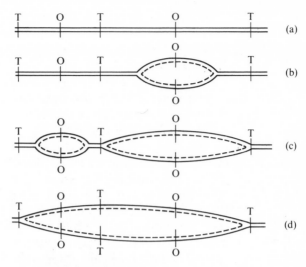

T O T O T   (a)

O

T O T   T   (b)

O

O   O

T   T   T   (c)

O   O

O T O   (d)

T   T

O T O

O — replication initiation site
T — replication terminus

FIG. 13. A bidirectional model for DNA replication in eucaryote chromosomes. The four figures ((a)–(d)) represent successive stages in the replication of two adjacent replicons. The horizontal lines in (a) are the two polynucleotide strands of a single DNA dimer. The interrupted lines in (b)–(d) are newly synthesized polynucleotide stands. O = replication initiation site, T = replication terminus (after Hubermann and Riggs (1968)).

(X/A) ratio is 5 per cent—the so-called simplex type. Here one of the two X-chromosomes present in female somatic cells is heteropycnotic and forms a prominent sex-chromatin body or Barr body at somatic inter-phase. This condition consitutes one of the clearest cases of facultive heterochromatinization. In common with other types of facultative heterochromatin the heterochromatic X is also late-labelling over most of its length.

The pattern of chromosome replication we arrive at, therefore, is one of long DNA molecules each containing a very large number of replicating units. These units are individually quite short so that each of them completes replication in a fraction of the total DNA synthesis or S period. Throughout S there is a set sequence of replication which is genetically determined. Why the property of facultative heterochromatinization should cause a temporal alteration in the control of DNA synthesis is not known, but it implies that the structural arrangement of nucleo-protein affects the initiation and rate of synthesis.

These remarks apply principally to mitotic cells but meiocytes appear to behave in an essentially equivalent manner with two qualifications.

(*i*) In the newt, meiotic DNA synthesis has been shown to be slower over-all, and this appears to result from the replicating units being much longer than their equivalents in somatic cells. One consequence of this is that replication is not completed until prophase of the first meiotic division.

(*ii*) There is also evidence for the occurrence of a quantitatively minor but qualitatively important repair synthesis during the mid stages of first prophase. The significance of this supplementary synthesis is discussed later (see p. 57).

2.2. The dividing nucleus

In bacteria where there is a single, relatively short linkage group, nuclear division is accomplished by a simple mechanism which appears to involve the plasma membrane. After replication the two halves of the replicated DNA strand have separate connections with the membrane. As the cytoplasm divides, these two points are pulled or pushed apart carrying the replicated units into separate daughter cells. Moreover, since the procaryotic cell has only one chromosome, only a 'like' chromosome can be produced by division. In *E. coli* and, presumably, in procaryotes generally, the single chromsome is attached to the cell membrane in the region of the replicating fork. There may also be a second point of attachment at the site of replication initiation. Growth of the cell membrane between these attachments appears to be the main mechanism for separating the products of chromosome duplication.

The mitotic mechanism found in eucaryotes resembles the process of cell fission in bacteria in that it too duplicates the genotype of the original cell in an exact manner so that two daughter cells are produced which are genetically alike unless mutation occurs during or after the process. By contrast, mitosis is a distributive mechanism which provides a basis for the equal distribution of a replicated genome of much greater complexity. During interphase, the period between division, and hence during replication, the chromosomes are enclosed by a membrane. Subsequent separation of the replication products is a function of a specialized system of protein microtubules which proliferate at a particular time in the cell cycle (prophase) and are then organized into a spindle-shaped apparatus on which the chromosomes orient (metaphase) and subsequently move (anaphase). In some of the simpler eucaryotes (Ciliophora, Protozoa, and Phycomycete fungi), the spindle is an intra-nuclear entity, but in most, the production of the spindle is correlated with the disruption of the nuclear membrane. Spindle microtubules are generated in the perinuclear cytoplasm immediately prior to the breakdown of the membrane. Indeed the microtubules themselves may play a part in inducing this breakdown.

The spindle system initially contains microtubule material of two kinds: first, polymerized aggregated units which run in a continuous manner between the two poles of the spindle (continuous spindle fibres);

and secondly, unpolymerized units, which occur in a disaggregated state. Orientations of the chromosomes within the continuous fibre system depend on assembly of these disaggregated units into specialized chromosome spindle fibres. These establish a temporary but firm attachment with each of the two chromatids in a chromosome.

In many eucaryotes the site of attachment of the microtubules is localized at the centromere region. And the localization of satellite DNA in the vicinity of this region raises the interesting possibility that it may play a role in the kinetic functions conventionally ascribed to this region. In some eucaryotes, however, spindle microtubules become attached at multiple sites on each chromatid (see p. 19). The precise mode of microtubule attachment in fact regulates the movement sequences which chromosomes undergo during division and so determines the basis of chromosome mechanics.

2.2.1. *Chromosome mechanics*

(a) *Mitotic manoeuvres.* Mitosis is essentially a simple sequence (Fig. 14) prior to which each chromosome duplicates to give a two-stranded structure, the component units of which are identical in morphology and genetic organization, and are termed chromatids. The self-replicating properties of DNA underlie this accurate duplication, and mitosis is simply a device to ensure the exact qualitative and quantitative distribution of the products of chromosome duplication. This is achieved by a regular and precise sequence of mechanical movements on the part of the chromosomes which result in the sister chromatids of each chromosome separating to opposite spindle poles. Why they do so is not clear. Certainly it does not depend on any geometrical property of the centromere face, since at initial attachment to the spindle mal-orientation is not uncommon. In such cases, however, reorientation occurs to ensure that the two sister chromatids engage to opposite poles in a state of auto-orientation. Once correctly oriented, the system is remarkably stable, so that if centromeres are disengaged from the spindle by micromanipulation, they rapidly reform an attachment toward the pole they are facing, irrespective to its proximity.

Orientation itself is not sufficient to guarantee efficient separation. Following successful orientation, each chromatid pair moves rapidly and irregularly between the poles until they all take up positions in an equatorial plane midway between the two poles. Here they may continue to exhibit minor oscillatory movements and remain balanced in such a position for periods of time ranging from a few minutes to a few hours, depending on the cell type involved. Such balance depends on two principal factors. First, the pull of the two microtubule systems on each sister centromere pair. This pull is in fact sufficient actually to separate sister chromatids in the region of the centromere. Secondly, each sister chromatid pair is maintained in intimate association. The

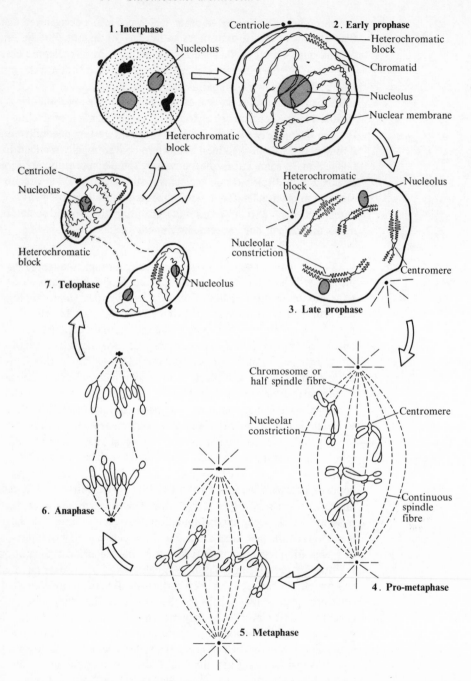

FIG. 14. The mitotic cycle.

nature of this association has never been defined but effective separation involves the progressive disassociation of one sister chromatid from the other, beginning at the centromere and extending progressively to the ends. Added to this, each chromosome spindle fibre shortens in conjunction with the disassociation of chromatids so that each chromatid moves at nearly constant velocity towards that pole to which it is oriented.

Mitotic movement is without question a consequence of microtubule activity and it is now generally assumed that linear movement results from an interaction between chromosome spindle fibres and the continuous spindle fibres. For example, it has been suggested that the microtubules attached to the centromeres are able to slide actively past the continuous microtubules in a directional and polarized manner. As each microtubule with its associated chromatid moves towards its pole the components of the leading end progressively disassociate. This, rather than contraction, leads to a shortening of fibre length, for the diameter of the fibre does not change.

A much simplified system, somewhat reminiscent of that found in bacteria, characterizes the hypermastigid flagellates. The chromosomes in these protozoans are permanently condensed and they remain enclosed in a nuclear envelope throughout the life cycle. An extranuclear spindle is developed in conjunction with two very specialized centrioles which act simultaneously as basal bodies for the organization of flagella. Sister chromatids are individually and terminally attached to separate sites on the inside of the envelope. Spindle microtubules attach at corresponding sites outside the membrane in such a manner that sister chromatids are drawn to opposite poles. Division is completed by elongation and subsequent constriction of the nucleus (Fig. 15).

(b) *Meiotic segregation.* Meiosis is a mechanism which, in conjunction with fertilization, regulates, amongst other things, the chromosome number of an organism. A second feature of the meiotic mechanism is a redistribution of genetic material into daughter nuclei which is made possible by the segregational and recombinational phases of the sequence. The basis for all the events of meiosis depends, in part, on the structural organization of the chromosomes and, in part, on the operation of specific gene systems which they themselves contain. Between them these two factors, structural and genotypic, regulate the pattern of chromosome mechanics and the synthetic capacities of the dividing nucleus.

The simplicity of mitotic manoeuvres is in marked contrast to the complexity of the movements which chromosomes undergo at meiosis. Two principal factors account for this difference:

(*i*) in mitosis each chromosome contracts and moves quite indepen-

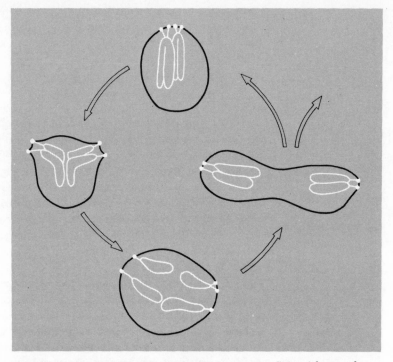

FIG. 15. Intranuclear mitosis in the flagellate protozoan, *Spirotrichonympha.*
The chromosomes are permanently condensed and attached to a persistent nuclear
envelope (after Cleveland (1949)).

dently of all the others irrespective of their homologies. During the
first division of meiosis, on the other hand, homologues behave in
concert and chromatids are associated in novel and complex patterns;
(*ii*) during the first of the two division sequences that characterize
meiosis, the centromeres show a distinct and different form of
orientation.
Let us examine chromosome behaviour at meiosis in these terms.

(*i*) *Chromatid associations.* At the onset of meiosis (Fig. 16) as in
mitosis, homologues are separate entities. By contrast to mitotic prophase chromosomes, however, they appear to be single (leptotene). An
active and specific pairing process then leads to the formation of a
haploid number of bivalents (zygotene). Each bivalent consists of a pair
of homologous chromosomes, and each of these chromosomes consists
of a single chromatid. The apparent singleness of each homologue may
reflect the fact that while the bulk of DNA synthesis is carried out
during the pre-meiotic S-phase, a particular fraction is not replicated
until zygotene. This DNA has a distinctive base composition and
although it has a high bouyant density it is not rDNA (see p. 66). What

is more, the splitting of a chromosome into two chromatids must involve more than mere DNA synthesis as is clear from the development of polytene chromosomes. Other syntheses and, especially, molecular reorganization are clearly required.

The mechanism by which pairing of homologues is achieved is not known but it is an attractive possibility that the DNA synthesized at zygotene might also serve to demarcate pairing initiation sites.

Chromosome contraction which is initiated at the onset of meiosis is often particularly pronounced following pairing and leads to the production of much shortened and thickened bivalents (pachytene). Homologues are still visibly single at pachytene, and it is not until the paired homologues begin the separation process which initiates the opening out of each bivalent that homologues can be seen to be double (diplotene). Each chromosome now consists of two chromatids. By this stage, in a majority of organisms, at one or more sites in each bivalent two of the four chromatids lie criss-crossed over one another at so-called chiasma points. On either side of each chiasma, sister chromatids lie in close parallel alignment. There is now compelling evidence that each of these criss-cross arrangements results from an X-type exchange between non-sister chromatids in each bivalent. Where homologues differ in allelic composition, such an exchange serves as a mechanism for recombining their differences, an event which in genetic terms is referred to as crossing-over (Fig. 17).

Shortly after the appearance of chiasmata, sister chromatid pairs on each side of each chiasma rotate relative to one another and the intervening chiasma. The net result of this is to transform the criss-cross alignments of the chromatids into a series of open crosses (diakinesis). Where only one chiasma is present in a bivalent, this has the effect of converting the bivalent into a cross-shaped configuration in one plane. Where two chiasmata are formed on opposite sides of the centromere, a ringe bivalent results. And where three or more chiasmata are formed per bivalent, multiple loops are formed (Fig. 18). Irrespective of any genetical function it may serve, a chiasma is always effective in maintaining an association between the homologues in a given bivalent, by virtue of the fact that sister chromatids continue to maintain a close parallel alignment in all regions other than at the chiasma, for chromatids, especially when contracted, are not infinitely flexible. Were it not for the chiasmata, therefore, the component homologues in each bivalent would disassociate when sister chromatid pairs tend to separate following pachytene.

The parallel alignment of sister chromatid pairs continues through the first prophase of meiosis until the bivalents in a nucleus have completed their orientation and congression on the first division spindle (metaphase-I). After this, sister chromatid association lapses and remains

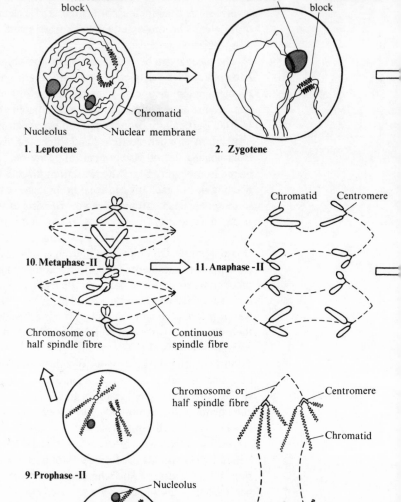

FIG. 16. The meiotic cycle.

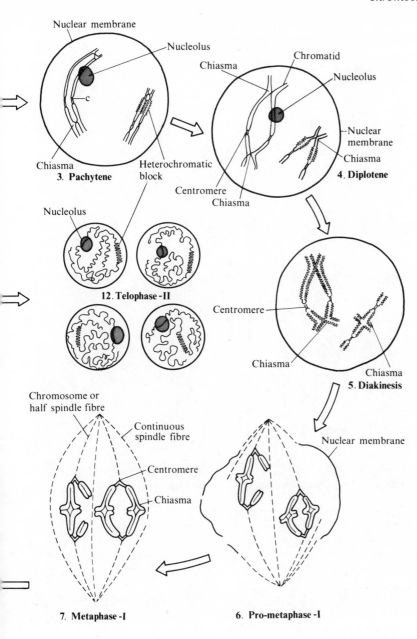

Nuclear membrane

Nucleolus

Chiasma

c

Chiasma

3. Pachytene

Heterochromatic block

Chiasma

Chromatid

Nucleolus

Nuclear membrane

Chiasma

Centromere

Chiasma

4. Diplotene

Nucleolus

12. Telophase - II

Centromere

Chiasma

Chiasma

5. Diakinesis

Chromosome or half spindle fibre

Continuous spindle fibre

Centromere

Chiasma

Nuclear membrane

7. Metaphase - I

6. Pro-metaphase - I

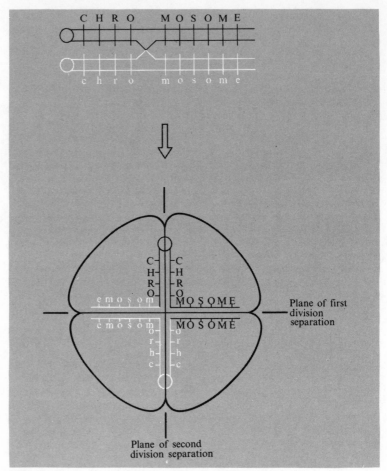

FIG. 17. The genetic consequences of chiasma formation.

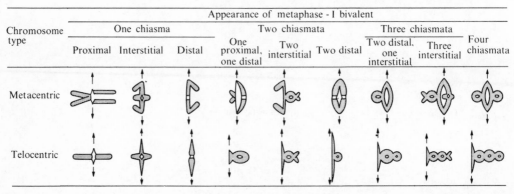

Chromosome type	Appearance of metaphase - I bivalent								
	One chiasma			Two chiasmata			Three chiasmata		Four chiasmata
	Proximal	Interstitial	Distal	One proximal, one distal	Two interstitial	Two distal	Two distal, one interstitial	Three interstitial	
Metacentric									
Telocentric									

FIG. 18. Bivalent morphology in relation to chiasma frequency and chiasma position.

lapsed through the remainder of the meiotic sequence except at relatively small procentric regions where they remain firmly connected. The nature of this association between sister chromatids is not known, but it is critical to the meiotic mechanism.

(*ii*) *Centric behaviour at meiosis.* Each meiotic bivalent has two pairs of sister centromeres, one of paternal, the other of maternal origin. At first division orientation, each sister pair makes an independent microtubular association within the spindle of continuous fibres. Sometimes both centromere pairs orient to the same pole, more usually they do so to opposite poles. Only the latter arrangement is stable and the former rapidly re-orients into a stable system. Following this co-orientation, bivalents congress into a metaphase-I alignment in which sister centromere pairs lie equidistantly above and below the spindle equator. Sister chromatid association lapses only when this stable position has been achieved. This lapse, coupled with microtubule-regulated centromere movement, leads to the resolution of each bivalent into its component half-bivalents. Each half-bivalent consists of two homologous chromatids and, in most meiotic systems, one or both of these have been subjected to one or more recombinational events.

Anaphase-I separation of half-bivalents produces two nuclei. The component chromatids of each half bivalent within a given nucleus are then separated from one another by a second division sequence in which the individual sister centromere pairs in each half-bivalent auto-orient in a manner reminiscent of mitosis. In this division, homologous chromatids remain widely splayed, reflecting the lapse of attraction which precipitated first anaphase. Completion of meiosis gives rise to four nuclei each containing a haploid number of chromosomes, each chromosome consisting of one chromatid.

The direction of co-orientation of the two sister centromere pairs in a given bivalent is usually random relative to that of the centromere pairs of all other bivalents. Consequently, maternal and paternal pairs segregate randomly at first anaphase. Likewise, the auto-orientation of half-bivalents at the second meiotic sequence is also a random affair. In cases where the maternal and paternal homologues differ in the alleles they carry, the net result of these two successive separation sequences is two fold:

(1) only one of the four chromatids of each four-stranded bivalent passes into a given meiotic product;

(2) centromeres, and hence the chromatids attached to them, segregate randomly and so there is a recombination of unlinked genes in case where genetic heterogeneity occurs between the various paternal and maternal members.

Of course orientation and separation mechanisms can fail, leading to a non-disjunction of homologous elements. In general, however, this is a rare event and it constitutes a form of numerical mutation (see p. 101). In *Drosophila* it has long been known that the presence of a Y-chromosome in the female greatly increases the frequency of non-disjunction of the X-chromosome. More recently Grell (1967) has described a number of cases where non-homologous chromosomes can show predictable inter- actions leading to anomalous segregation behaviour. Thus females carry- ing a Y-chromosome and a single free chromosome IV show Y-IV segre- gation in 92 per cent of the meiotic products instead of the 50 per cent Y-IV distribution predicted on the basis of the random movement of unassociated univalents. Grell (1967) also has evidence for meiotic interactions leading to segregation between (a) Y + II, (b) Y + III, (c) X + IV, (d) X + II, (e) Y and 3:4 translocation, (f) X and a 3:4 trans- location, (g) X and a 2:3 translocation, and (h) X and an X-duplication. In all these cases the segregation frequencies show a highly significant departure from the 50 per cent expected with independent assortment.

The basis of this preferential segregation is not known. The most obvious suggestion to account for the segregation of apparent non- homologues is that these elements do have some regions of shared homology of an unusual kind which functions to secure segregation in the absence of conventional homologous association. An obvious candidate for such regions would be repeated sequence DNA. Non- specific associations have long been known to occur commonly between heterochromatic regions, even in the presence of homologous pairing, though they usually lapse prior to orientation in cases where chiasmata are formed. And some types of constitutive heterochromatin are known to contain repeated sequence DNA (see p. 21).

The balance of evidence in *Drosophila,* however, is that the critical factor in securing preferential segregation of predominantly non- homologous entities is either chromosome size or else some property closely related to size. Grell (1967) constructed a karyotype in which three modified chromosomes were introduced together into the same nucleus. Two of these carried the same centromere and the same heterochromatic component but had euchromatin of different origin and length. The third had no known homology with either of the other two but resembled the larger one in size. It was found that the two larger chromosomes segregated 98 per cent of the time while the small chromosome, though sharing proximal homology with one of the others, assorted randomly relative to both.

There is a number of other cases where particular chromosomes are regularly recovered in excess of 50 per cent from given hetero- or hemi- zygotes. Such systems of meiotic drive are of two kinds, the one operative in females, the other in males.

1. Female drive systems depend on the preferential segregation of a particular chromosome to a particular spindle pole at meiosis. Only one,

and frequently only a particular one, of the four products of each female meiosis produces a functional egg. Consequently any chromosome which is included preferentially in this nucleus will be subject to drive. For example, in the moth *Talaeoporia tubulosa* the female is XO and the male XX in sex-chromosome constitution. The univalent X of the female may move either towards the central pole or the peripheral pole of the first division spindle. In the former event an X-egg is produced which, on fertilization, gives an XX male. In the latter case a nullo-X-egg forms, the X passing into the non-functional polar nucleus. This, on fertilization, gives an XO female.

The movement of the X becomes markedly non-random at particular temperatures or at a particular age of the egg. Thus at 35 °C or in overripe eggs there is a preferential recovery of X-eggs, while at 10 °C or in young eggs there is a preferential recovery of nullo-X-eggs. This allows for an adaptable sex ratio which is particularly important in this species because the female is immobile.

An equivalent behaviour is known also for some univalent supernumerary or B-chromosomes at female meiosis. These too pass preferentially to the functional egg nucleus as a consequence of preferential disjunction.

2. Male drive systems are different in kind. Normally all four products of male meiosis are functional. Here, therefore, drive is only possible given a loss of certain of the meiotic products. For example, in the male of *Drosophila melanogaster* a dominant genetic factor SD (segregation distorter) is known where males heterozygous for this factor (SD^+/SD) regularly give progenies in which nearly all the offspring carry the dominant mutant SD. The SD locus is located near the centromere of chromosome II. Females heterozygous for the SD locus show no evidence of distortion.

Meiosis in SD^+/SD males is normal but, as Peacock and Miklos (1973) point out, SD^+ spermatids fail to complete spermiogenesis so that no functional SD^+ sperm form. The basis for such a failure is not yet known but clearly drive here does not depend on preferential segregation. Rather it depends on sperm dysfunction involving a breakdown of development in spermatids carrying a particular chromosome.

Finally, there are cases where chiasma formation is dispensed with either in particular chromosomes or else throughout the entire genome. The best example of the former is the sex-chromosome system in certain animals. Here segregation in the heterogametic sex is achieved either by a specialized form of pairing which does not depend on conventional homology or else without any form of pairing at all. In the latter case all homologues pair and segregate without forming chiasmata. In consequence there are no conventional diplotene or diakinesis stages in meiosis. Rather pachytene bivalents pass directly to pro-metaphase I.

Such achiasmate meioses are known in the males of all higher Diptera including *Drosophila*. They are also found in some individuals of the following groups—Protozoa, Oligochaeta, molluscs, mantids, acarines, copepods, scorpions, grasshoppers, bugs, and butterflies.

2.2.2. *Synthetic capacities*

(a) *Formation of the spindle.* Spindle microtubules are composed of small protein micelles. Indeed most of the spindle can be accounted for by one kind of protein which has a molecular weight of 34 700 and a sedimentation coefficient of 2·5S and which occurs both as a 3·5S dimer and a 22S polymer. This protein appears to correspond to the 33 Å sub-units seen in electronmicrographs of spindle microtubules.

Since spindle microtubules are constructed from protein, their components must be synthesized at ribosome sites by specific mRNA sequences under the control of genetic DNA. But whether this DNA is of nuclear or cytoplasmic origin has never been clarified. In most animals a pair of specialized organelles called centrioles are present which are associated with the spindle system. A centriolar pair is present where spindle microtubules converge at each pole. Each centriole has nine triads of microtubules as component parts of their own organization.

There is some suggestive, though indirect, evidence involving centrioles with the three major known microtubular systems of eucaryote cells, namely, cilia and flagella, cytoplasmic microtubules and the spindle microtubules. The most consistent relationship is that between centrioles, and cilia and flagella. Both of these contain nine sets of doublet microtubules arranged around the circumference of a circle, plus two larger ones lying at the centre (9 + 2 arrangement). These nine doublets are structurally continuous with the nine triads of microtubules found in the centriole. Thus the centriole appears to organize the assembly of the protein monomers that form the 9 + 2 tubules and there is the possibility that it mediates in their synthesis since centrioles appear to contain DNA. Moreover, the centriolar pair present at interphase normally begins to duplicate at about the time that nuclear DNA replication begins and its completion more or less coincides with that of DNA synthesis.

At prophase each pair of centrioles becomes the centre of a radiating system of microtubules known as an aster. During prophase the two centriolar pairs separate and the cytoplasm between them contains oriented microtubules that appear to represent the beginning of a new spindle. By the time the spindle is fully formed a centriolar pair is present where the microtubules converge at each pole. Notice, however, that spindle microtubules are not in contact with the centrioles since they end in a zone some several hundred ångströms away.

In most higher plants, on the other hand, centrioles are not represented in the cell structure; significantly neither are flagella or cilia. How the spindle microtubules are organized in this case is not clear. Whether there are self-replicating cytoplasmic entities which serve to template the mRNA needed for their synthesis, or whether this is a function of chromosomal mDNA, is not known.

Since multiple microtubules become attached to the centromere, and only to the centromere, of each chromatid, it appears that chromosomal DNA can certainly bind to spindle protein. That such binding may itself involve some synthetic activity on the part of the centromere is suggested by the fact that the centromere remains relatively uncoiled at a time when the rest of the chromosome is compact and condensed.

(b) *The chemistry of cross-over.* Crossing-over, the exchange event leading to chiasma formation at meiosis shows five basic characteristics:

(*i*) it occurs at an effectively four-strand stage when each homologue either already consists of, or else is being converted into, two chromatids;

(*ii*) only two of the four chromatids in the bivalent are involved in any given crossover and these are non-sisters of distinct parental origin;

(*iii*) with rare exceptions, the exchange produces precisely reciprocal products;

(*iv*) when very short segments are considered, the frequency of apparent recombination in them proves to be much higher than would be expected on the basis of their length (negative interference); and

(*v*) cross-over frequently can be shown to be a polarized phenomenon with clear directional properties.

In terms of chromosome organization as we have discussed it, crossing-over at the molecular level presumably occurs between paired nucleotide bases. Indeed, recombination must occur with considerable molecular precision so that nucleotides are rarely omitted or duplicated, since this would result in an alteration of the nucleotide code. It is not known how such fine pairing is achieved in chromosomes that are relatively condensed when pairing begins and which usually become more compact as the meiotic prophase proceeds. Certainly the synaptic association of homologues does not appear to provide for such intimate pairing. How then is it achieved?

At zygotene, when homologous chromosomes pair with one another, a highly organized and co-planar set of protein strands, the synaptonemal complex (SC), develops as a tripartite ribbon between each pair of homologues (Fig. 19). The lateral dimension of this structure, which ranges from 1600–2000 Å, places it just below the limit of resolution of the light microscope. At the time the SC develops, the two ends of each chromosome are attached to the inner surface of the nuclear membrane.

Chromosome ends sometimes contain repeated base sequences which

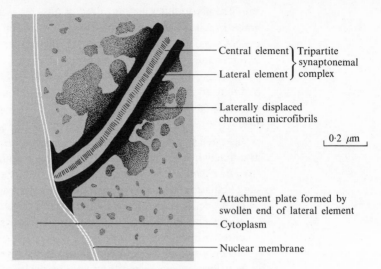

Central element ⎫ Tripartite
            ⎬ synaptonemal
Lateral element ⎭ complex

Laterally displaced
chromatin microfibrils

0·2 μm

Attachment plate formed by
swollen end of lateral element

Cytoplasm

Nuclear membrane

FIG. 19. General structure of the synaptonemal complex.

may facilitate initial pairing. If this is the case there must be some basis
for distinguishing homologous from non-homologous chromosomes.
Genetically different viruses are known to attach to different specific
sites on the surface of their hosts, and equivalent sites, specific for
particular chromosome ends or telomeres, may occur on the inner side
of the nuclear membrane.

While the SC appears to provide a mechanism for the crude alignment
of homologues, it cannot itself afford a basis for holding DNA molecules in the correct spatial alignment for exchange. The SC may,
however, provide a rigid structure within or around which recombination can occur. That is the SC may in some manner facilitate DNA
exchange. However, it is clear that synaptonemal complexes are not
sufficient to ensure exchange because they have been described in the
meiotic cells of organisms which do not indulge in crossing-over.

It is now generally recognized that the reconstructed DNA molecules
produced by the process of exchange are not, or need not be formed by
an end-to-end association of polynucleotide columns which were broken
at exactly reciprocal sites. There is rather a lateral association of complementary single-stranded segments which gives rise to sections of
hybrid DNA. This process allows for the correction or repair of any mismatched pair within the hybrid region so that normal base pairing is
restored. Significantly, the second and quite distinct S-phase which
occurs at pachytene appears to be concerned with repair replication
because hybridization studies do not reveal any net DNA increase
during this phase. By implication, therefore, these new DNA segments
presumably replace existing ones. This pachytene DNA varies in base

composition even within a species, and its average composition approximates to that of the total nuclear DNA (Table 9).

Table 9
DNA synthesis in pollen mother cells of Lilium

Stage	Molar base composition of $[^{32}P]$ DNA			
	C	G	T	A
1. Pre-meiotic S	20	22·8	30·2	27·0
2. Prophase S				
(a) Zygotene-S	26·1	27·0	25·0	22·7
(b) Pachytene-S	25·3	25·3	25·3	24·1
3. Total DNA	20·2	21·4	29·6	28·9

Note. After Hotta and Stern (1971)

A molecular mechanism for crossing-over can be visualized on the basis of the data currently available. The DNA binding protein discovered by Hotta and Stern (1971) may serve to bring two non-sister homologous chromatids into the precise alignment required for recombination. An endonuclease then causes single strand breaks in DNA. In a study based on microsporocytes of *Lilium* such an endonuclease was found to produce breaks leading to the formation of 3'-phosphoryl and 5'-hydroxyl termini at the breakage sites. These sites may need modification by phosphatase or 3', 5'-exonuclease to stimulate repair replication since polynucleotide ligase, which acts subsequently in the sequence to re-establish covalent linkages in the DNA backbone, is known to require 3'-hydroxyl and 5'-phosphoryl termini in juxtaposition in order to function. Additional DNA is then synthesized at pachytene to fill any gaps and, finally, polynucleotide ligase is required to integrate the newly replicated fragments of DNA (Fig. 20).

This model has the virtue of being able to accommodate most of the basic characteristics of crossing-over, including gene conversion, polarized recombination, and negative interference.

(*i*) *Gene conversion*. On rare occasions, allelic differences may segregate to give 1:3 or 3:1 ratios instead of the expected 2:2. Moreover, in ascomycetes, where the four nuclei produced by meiosis go through a post-meiotic mitosis before cell walls are laid down to give eight ascospores, conversion asci may give not only 6:2 or 2:6 ratios but, in addition, 5:3, 3:5, or even 1:7 and 7:1. These various aberrant ratios, and hence gene conversion, can be explained on the basis of the extent and direction of correction of mismatched bases in heteroduplex DNA regions.

(*ii*) *Polarity*. If the endonuclease producing the first break is site-specific then segments of DNA will be opened up for recombination only in particular directions from particular points. The predomin-

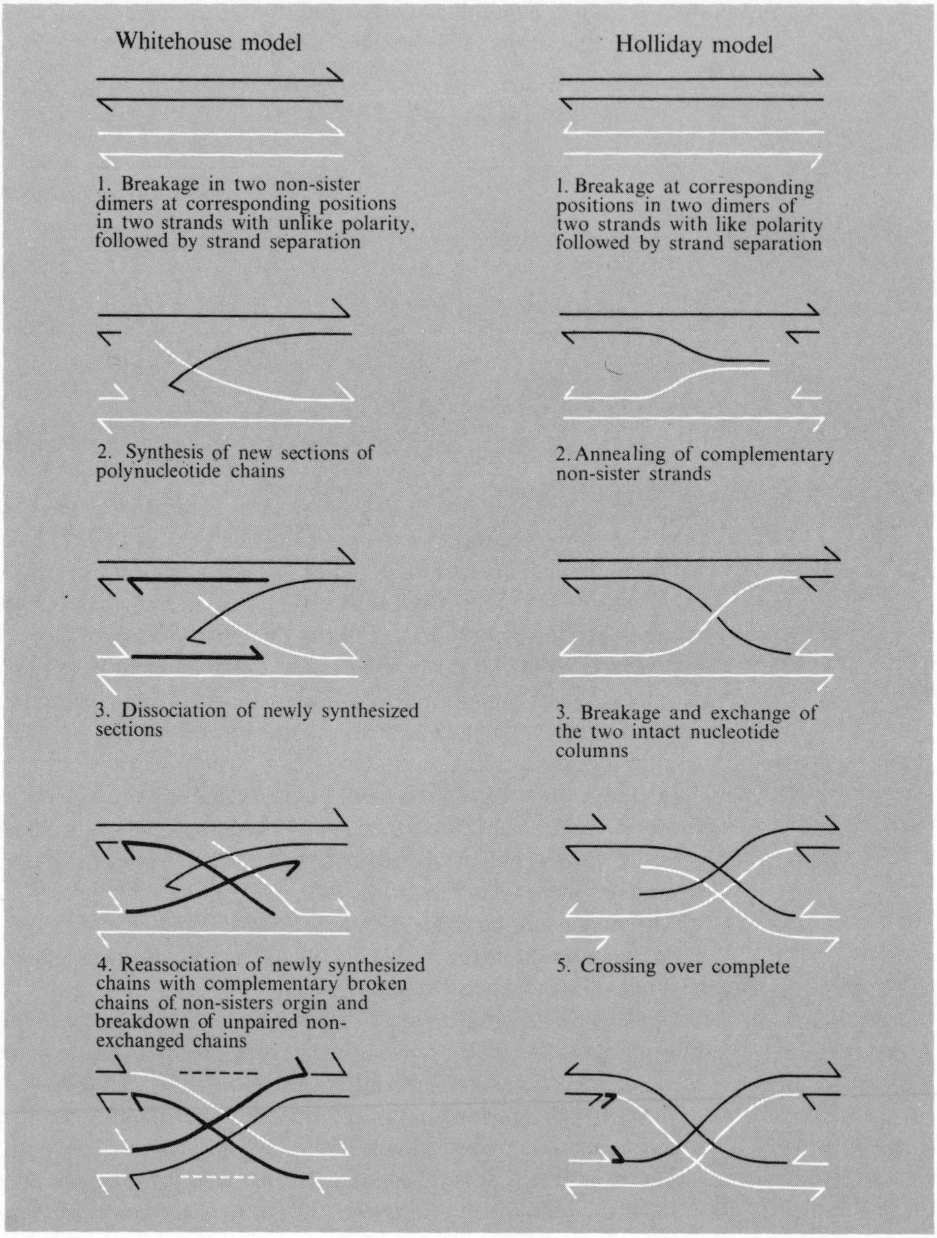

Whitehouse model

1. Breakage in two non-sister dimers at corresponding positions in two strands with unlike polarity, followed by strand separation

2. Synthesis of new sections of polynucleotide chains

3. Dissociation of newly synthesized sections

4. Reassociation of newly synthesized chains with complementary broken chains of non-sisters orgin and breakdown of unpaired non-exchanged chains

Holliday model

1. Breakage at corresponding positions in two dimers of two strands with like polarity followed by strand separation

2. Annealing of complementary non-sister strands

3. Breakage and exchange of the two intact nucleotide columns

5. Crossing over complete

FIG. 20. Molecular models of crossing-over. Note: only two of the four chromatids of a meiotic bivalent are shown and each is assumed to consist of a single DNA dimer.

ant direction of polarity would then indicate at which end a particular segment is opened up.

(*iii*) *Negative interference.* If two mutational differences at distinct but nearby sites within the same crossover segment are corrected at the same time by the excision-repair system then two apparent exchanges would occur within a short region and so lead to a correlated negative interference.

3 *Epigenetic activities*

The simplest conception of a chromosome is that of a string of contiguous cistrons variously influenced by factors controlling transcription. On this model, chromosome replication would be expected to produce two identical copies of the cistron string which are then distributed equally at cell division into the two daughter cells. With one added complication the gene string of procaryotes appears to conform to this simple model, or at least to approach it very closely. The complication is provided by the presence of systems for the co-ordinate control of polycistronic segments, though even in bacteria, such operons, as they are called, are scarcely ubiquitous. An operon is composed of various complementary parts—structural genes which transcribe mRNA, regulator genes, and an enhancer or promoter site. The promoter is the region of the operon at which initiation of transcription is controlled. Such control involves the recognition and binding of RNA polymerase to DNA and the opening of the duplex. The structural genes, each of which codes for a particular polypeptide and whose enzymic products are concerned in the same biochemical pathway, form a contiguous series along the chromosome (Fig. 21).

This cluster of functionally related cistrons is co-ordinately controlled by interaction between the product of a specific regulatory gene and an operator region which is located at one end of the cluster. However, the regulatory gene need not adjoin the cistronic cluster it controls and it can lie anywhere in the genome. It codes for a repressor substance that inhibits operon function. The repressor binds specifically to the promoter and so prevents either the attachment or the function of RNA polymerase. Gene clustering of this kind is not present in eucaryotes. Even when some clustering is found as in the ribosomal cistrons (see p. 19), there is no evidence for an associated operator site. In fact the situation is generally much more complicated and enzyme systems that are co-ordinately controlled in procaryotes are generally uncoupled in eucaryotic organisms. Consequently, differential gene function, both in time and place, is an intrinsic and fundamental aspect of genetic and epigenetic activity.

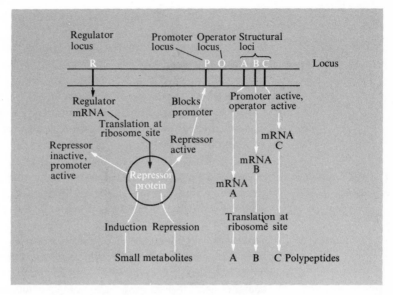

FIG. 21. The organization and regulation of an operon.

In eucaryotes, development depends on a carefully controlled
sequence of genetically determined processes. Essentially, therefore,
epigenesis implies the regulation of effective gene activity. Three
direct levels at which such regulation could be effected are as follows.

(a) *Transcriptional control.* One or more proteins of the chromosome,
or else a species of RNA, may specifically mask a particular segment
of DNA and so prevent its transcription.

(b) *Translational control.* Messenger RNA must be packaged protec-
tively prior to its passage from the intranuclear sites of transcription
to the extranuclear sites of translation. Otherwise it would be
degraded rapidly by nucleases, which abound on both sides of the
nuclear membrane. But protection from destruction also confers
concealment from polymerase and other components of translation.
Hence manipulation of the mechanisms of unmasking provides a
point for the post-transcriptional control of effective gene action.

(c) *Post-translational control.* Many functional proteins do not
consist simply of single polypeptides even in their tertiary form.
Rather they are multimeric aggregations of similar (homomultimeric)
or dissimilar (heteromultimeric) polypeptides or a combination of
the two. The ultimate activity of genes could therefore be regulated
by manipulating the processes of aggregation and, hence, quaternary
structure. What is more, the activity of even structurally functional
enzymes is limited by a variety of other factors.

Apart from these simple systems, however, eucaryotes possess more widespread methods for regulating gene expression, involving differential gene content. As a consequence of differential replication, inactivation or elimination of a part of the genome, the number of copies of a given gene, or the number of genes capable of activity may differ in different cells or even in the same cell at different times. Thus, whereas DNA is transmitted to the gametes in a constant amount, it is subject to a number of processes which modify its dosage and effectiveness at particular times in particular tissues. And all these mechanisms have a chromosomal basis.

3.1. Amplification mechanisms

In eucaryotes a particular cell type is often required to produce a large amount of a single protein. There are two principal ways in which such heightened production can be achieved.

3.1.1. Reiteration

This refers to the presence in all cells of multiple copies of a particular gene. For example, in *Drosophila melanogaster,* some 60 different kinds of gene are known to transcribe tRNAs and there are about 13 duplicated copies of each of these genes. The genes controlling histone production also appear to be highly reiterated.

From a theoretical point of view there are potential problems in incorporating multiple copies of the same gene into a genome, especially when these are organized as a tandem series. In the first place a defective mutation in one member of a reiterated series would presumably be sheltered against selection. To overcome this and other problems Callan (1972) has proposed that there is a strict hierarchy among tandem duplicates so that only one copy, the master, in any given series is replicated in preparation for transmission. Only if the master copy in any given series suffers a mutation will the slave members of that series also exhibit the mutation. In at least one known instance of a tandem series we can be confident that this explanation is insufficient. The case in point is the rRNA gene sequence found at the nucleolar organizer. Here the gene for 28S and 18S rRNA occur as tandemly reiterated sequences which certainly do not behave in a master-slave hierarchy. The master-slave hypothesis is also at variance with DNA renaturation studies which indicate that many essential eucaryotic genes must be present as single copies.

A further difficulty in a tandemly duplicated system is that unequal crossing-over is possible so that the system is inherently unstable (see p. 30).

3.1.2. Amplification

In contrast to reiteration, differential amplification involves the over-replication of particular genes in particular cells at particular times of development in relation to the production of large quantities of a particular component. One of the clearest cases of such amplification is the maternal programming which is involved in the formation of the oocyte in many animals. For example, in *Xenopus*, rRNA synthesis under the control of the embryo genotype does not begin until gastrulation, and even then its pace is slow. It is not until the tail-bud is forming that there is a significant increase in the net amount of rRNA and ribosomes. By contrast, in mammals, rRNA synthesis begins very early. In the mouse, for example, it starts following the second cleavage mitosis and it is at this time that nucleoli appear in the early blastomeres. In *Xenopus*, on the other hand, nucleolar development begins at gastrulation.

Throughout early embryogenesis a developing organism often depends predominantly or solely on the rRNA and mRNA synthesized during oogenesis which is then conserved for future use. To this end the growing oocytes of many animals remain in a prolonged phase during which the bivalents increase enormously in size. Coincident with this increase, a large number of laterally projecting loops develop at chromomere sites from each chromatid in the four-standed bivalent. For example, in *Triturus*, some 20 000 such loops are formed. These paired lateral loops become surrounded by an asymmetrically disposed loop matrix (Fig. 22). Labelling with [^{3}H] UdR or labelled amino acids shows that the overwhelming majority of lateral loops take up the isotope uniformly through their length. Moreover, inhibitors of RNA synthesis, such as calf-thymus histones (see p. 89) and actinomycin-D, cause the collapse of the loop system. Labelled RNA remains in the loops where it is synthesized for over 2 weeks and at least two-thirds of this informational RNA is then stored in the oocyte for the duration of oogenesis and is inherited by the embryo as the major component of a large stockpile of maternal mRNA.

As yet there is no reliable evidence as to whether each loop produces more than one species of gene product. Indeed it is assumed that each loop represents a single genetic locus. Callan believes that each loop represents the linear replicates of a single gene; that is, each loop contains repetitive DNA arranged functionally in a master-slave relationship. He has also suggested that the lampbrush system represents the means whereby the master copy vets and corrects the sequence in the slave replicates so that a mutation in the master is automatically transferred to the slaves. An alternative possibility is that the loop does indeed contain replicate genes which are all expressed and which serve indispensible functions possibly in relation to the stabilization of messenger molecules in preparation for their period of prolonged storage.

A second feature of the lampbrush oocyte is the immense number of nucleoli formed. In *Triturus*, for example, in excess of 1000 micronucleoli are formed while in *Xenopus* the comparable figure is 600–1200. Two general mechanisms appear to lead to such a multinucleolate con-

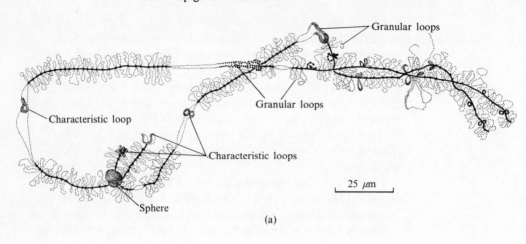

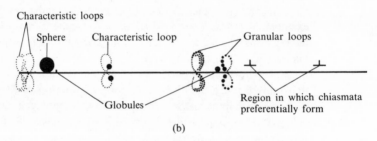

FIG. 22. Structure of lampbrush chromosome No. XI in oocytes of *Triturus vulgaris meridonalis:*
  (a) normal appearance;
  (b) cytological map (after Mancino *et al.* (1970)).

dition. In *Xenopus* and *Bufo* most of the extrachromosomal DNA is synthesized during the pachytene stage of meiosis and, as Gall (1969) has shown, its amplification can be followed in cytological preparations. As pachytene proceeds, Feulgen-positive granules or filaments accumulate in the immediate vicinity of the nucleolus. These eventually collect together opposite the polarized chromosome ends which form a conspicuous nuclear cap.

Beginning in late pachytene and continuing into early diplotene, the cap material spreads over the inside surface of the nuclear envelope. The leptotene nucleus usually contains a single nucleolus but additional nucleoli begin to appear during pachytene and especially during early diplotene. These multiple nucleoli are developed from DNA granules which are derived from the disaggregation of the nuclear cap material. DNA measurements confirm the synthesis of large quantities of extrachromosomal DNA. In conventional DNA cycles (Fig. 23) the cells

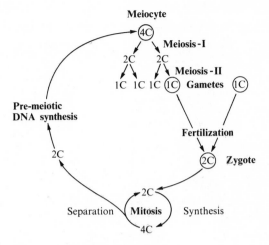

FIG. 23. The DNA cycle in Metazoan animals.

entering meiosis have a 4C-level of DNA but in *Xenopus,* where the
4C-DNA value is 12 pg, late pachytene oocytes contain some 43 pg
DNA. Thus there is nearly three times as much extrachromosomal
nuclear DNA as chromosomal DNA. In consequence we can estimate
that the oocyte now contains some 2500—5200 nucleolus organizers.
These extra copies are isolated in multiple micronucleoli which appear
to function only during oogenesis, serving as sources of rRNA which is
stored and conserved in the egg cytoplasm. They are not replicated
during cleavage and are in some way subsequently rendered non-
functional and ultimately discarded. The genes for 5S rRNA on the
other hand are not amplified at this time but they are already
highly redundant (see p. 19). The morphological events in *Bufo* are
similar to those described for *Xenopus* and here the 4C-value of the
early oocytes is raised to a maximum of about 10C in pachytene nuclei,
increasing the DNA content from 6·5 pg to 45 pg.

In plethodont salamanders and in the axolotl, on the other hand, the
extra peripheral nucleoli are produced by, and shed from, one or more
nucleolar organizers at the lampbrush stage. Here the nucleolar DNA is
represented by some 200—300 ring-shaped structures which vary greatly
in size. The rings give a positive Feulgen reaction and can be disrupted
by DNAase. On the other hand, RNAase and protease do not disrupt
them but these enzymes do remove nucleolar material. These nucleoli
thus appear to be circular molecules of DNA with associated RNP
complexes. In fact these rings can be shown to originate from the
nucleolar organizer which remains in an extended state long after other
loops have retracted. One further difference is that the nucleolar loops,
unlike other loops, do not arise as paired structures (Fig. 24).

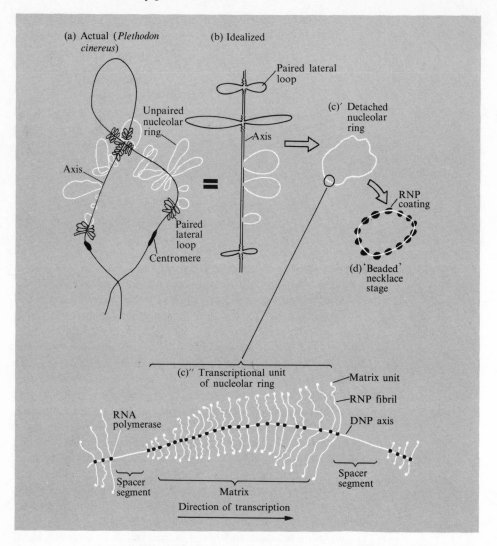

FIG. 24. Formation and structure of nucleolar rings in Plethodont salamanders.

The only difference between the extrachromosomal rDNA and that of the normal organizer is the absence of 5-methylcytosine in the former case. The significance of this difference is not known. May be the methylation of cytosine residues is of importance in the integration of rDNA into the chromosome.

There is convincing evidence that these extrachromosomal nucleoli function in the same manner as integrated organizers in the synthesis of rRNA precursors. Thus, the fibrous core of DNA which these nucleoli contain can be isolated and dispersed for electron microscope study.

Each unwound core can then be seen to consist of a thin fibre 100–300 Å in diameter which is periodically coated along its length with matrical material (Fig. 24). The axial fibre of each core forms a circle and treatment with DNAase breaks the core axis, The diameter of trypsin-treated axial fibres is about 30 Å which suggests that the core axis is a single Watson-Crick dimer coated with protein. The matrix segments along the core axis all show similar polarity and are separated from their neighbours by matrix-free axis segments forming a system of spacers. Each matrix unit consists of about 100 thin fibrils connected at one end of the core axis and varying in length from the thin to the thick end of the unit. Treatment with ribonuclease, trypsin, and pepsin all remove these fibrils. At the point of attachment of the RNP fibril to the axis there is a 125 Å granule. Its size and location are consistent with its being RNA polymerase. Miller believes that each matrix-covered DNA segment is in fact a single gene locus coding for rRNA precursors and in *Xenopus* cores from 8–175 units long are commonly found while in larger cores estimates of up to 1000 units have been recorded.

These observations on female amphibian meiosis indicate that there is a selective amplification of those DNA units which code predominantly for rRNA in the oocyte. Indeed some 97–99·5 per cent of the RNA synthesized in such oocytes can be shown to be ribosomal in character. Such selective gene amplification can be viewed as a mechanism for increasing the synthetic capacity of the oocyte without disrupting the normal course of meiosis. Apart from the selective amplification of ribosomal cistrons, informational mRNA is also synthesized during the lampbrush phase and then stockpiled together with the rRNA in anticipation of the requirements of cleavage and blastulation. In this connection, it will be appreciated that the tadpole does not become nutritionally independent until it reaches a comparatively late stage of development.

A comparable escalation of rRNA can be demonstrated in other oocytes too. For example, in the fly *Tipula oleracea,* a DNA body is formed in contact with sex chromosomes in oogonial interphase and is indistinguishable in appearance from heterochromatin. At each oogonial mitosis this DNA body passes undivided to one of the anaphase groups. It then increases appreciably in size during the pre-meiotic interphase. This body accounts for some 60 per cent of the nuclear DNA. Chromosomal DNA synthesis and that in the DNA body occur at different times, but both DNA fractions are complexed with histone.

Electron-microscope studies show that nucleoli occur inside the DNA body (Fig. 25), and a band of RNA is found between it and the chromatin. Since the DNA is complexed with histone, and nucleoli are situated inside the body, the simplest interpretation is that it represents hundreds of copies of the ribosomal genes produced by selective

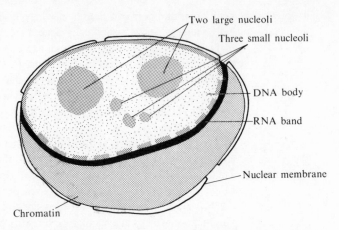

FIG. 25. Electron-microscope structure of the DNA body in *Tipula* (after Lima-de-Faria and Moses (1966)).

amplification of the nucleolar organizer. This DNA body disintegrates at late diplotene releasing its DNA.

Similar structures are known also in the oocytes of the beetle *Dytiscus marginalis* and the cricket *Acheta domestica*. Both these cases are again indicative of a differential amplification of the rDNA genes located in the nucleolus-associated heterochromatin. Indeed there is a growing body of evidence that such an amplification characterizes many oocytes. It is not, however, found in all organisms since the synthesis of RNA may be taken over partially or wholly by special nurse cells. In such cases, the nurse cell nuclei are highly endopolyploid and engaged in active RNA synthesis. This RNA is then transported to the oocyte by cytoplasmic bridges and there is no amplification of the rDNA genes of the oocyte itself. For example in insects with panoistic ovaries lampbrush chromosomes are present in the oocytes whereas in forms with meroistic ovaries there is no lampbrush phase and the oocyte RNA is synthesized by special nurse cells which are either polytene or endopolyploid. These nurse cells originate from the same germinal epithelium as the oogonium itself and are characteristically joined to it. In fact initially both the first nurse cell and the oocyte develop synaptonemal complexes but those of the nurse cell later disappear. A particularly interesting case has been described by Ribbert and Bier (1969) in *Calliphora erythrocephala*. Here the highly polytenized nurse-cell nuclei increase their template capacity for synthesizing rRNA in a manner similar to that of amphibian oocytes by producing multiple free nucleoli which synthesize RNP.

Equivalent nurse, or accessory cells are known also in the embryo sacs of angiosperms and again these are often either endopolyploid or polytene in character, though in the latter event they lack the chromatic banding found in dipteran polytene systems. Cells of the embryo suspensor of *Phaseolus* also appear to be capable of extra DNA synthesis not only in the nucleolus organizing regions but in several other areas too. This applies to endopolyploid cells of the suspensor 'handle' and the polytene cells of the suspensor 'knob'.

The clustering of a large number of 18S and 28S rRNA genes at the nucleolar organizer may, therefore, reflect both the demand for large quantities of their products and the requirement for further amplification at oogenesis. Moreover, the occurrence of augmented rDNA production suggests the need to distinguish between genetic DNA and metabolic DNA, the former being concerned with hereditary transmission, the latter with developmental regulation. Yet a further instance of this same phenomenon can be found within polytene systems.

We have seen earlier (p. 35) that the number of chromatid equivalents in each polytene chromosome depends on the number of replication cycles that occur during their development. Since there are two somatically paired homologues per potential polytene unit, then 8 cycles of replication give 512 units, 9 cycles give 1024 units, and so on. However, all regions do not undergo the same number of replications. For example, the centromeric heterochromatin of *Drosophila* either does not replicate at all or else goes through fewer replication cycles than euchromatic regions—it is under-replicated. Likewise, the heterochromatic arm of the X and the entire Y remain at the 4C level while the euchromatic regions of the same nucleus may reach a polytene level of 2048C. The Y-chromosome carries genes for rRNA while the nucleolar organizer in the X is also located within heterochromatin. Henig and Meer (1971) have recently found that the relative numbers of rRNA genes are distinctly different (Table 10) in DNA originating from diploid cells (fertilized egg) and highly polytene cells (salivary gland). Thus, in line with the lack of polytenization of heterochromatic material, there appears to be a relative under-replication of rRNA cistrons in polytene cells though they are certainly not at the 4C level characteristic of the heterochromatin itself.

Table 10

Representation of rRNA genes in egg and polytene salivary gland nuclei of D. Hydei

Cell type	% DNA—RNA hybridization	rRNA genes per haploid genome
1. Egg	0·39	280
2. SG nuclei	0·066	30

Note. Data from Hennig and Meer (1971).

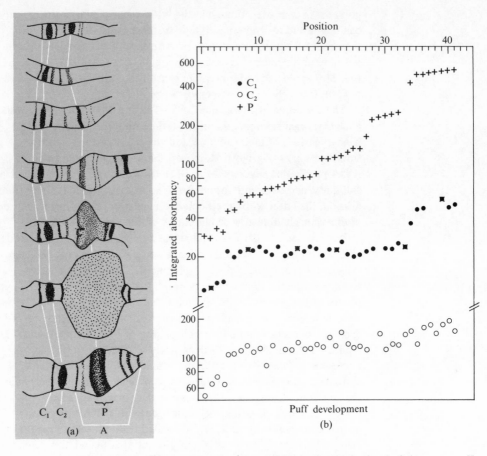

FIG. 26. Development of a DNA puff (A) in the proximal end of chromosome II in *Sciara coprophila.*

(a) Progressive development of puff A. The lines connecting successive stages mark the position of control bands C_1 and C_2 and band P in puff A.

(b) Semi-log plot of DNA increase in bands C_1, and C_2, and subunit P of puff A. The sets of measurements corresponding to the stages shown in (a) are marked x on the C_1 values (after Crouse and Keyl (1968)).

Conversely, differential over-replication is evident in the so-called DNA puffs which have been described in the *Sciaridae*. For example, in *Rhynchosciara angelae* one locus in section 2 of chromosome B swells enormously, accumulates DNA by extra replication, and returns to a banded state. Here puffing is indicative not of transcription but of extra replication and the extra DNA produced is retained by the band.

Sciara coprophila also shows localized increases in DNA (Fig. 26). Here they occur in at least four specific regions of the complement and begin just before pupation. By measuring the DNA content of these individual puffs, Swift (1961) showed that DNA augmentation is

involved here too. Comparable DNA-synthesizing puffs have not been
found in *Chironomus* or *Drosophila* but, since these have discrete
organized nucleoli whereas the Sciarids do not, Shultz has suggested
that DNA puffs represent an alternative mechanism for amplifying rDNA
in species where localized nucleoli do not occur. Like RNA puffs these
DNA puffs too may be asymmetrical and heterozygous. By [^{3}H] TdR
uptake, Perondini and Dessen (1969) have been able to show that in a
heterozygous DNA puff of *Sciara ocellaris* (band 9C) one homologue
produces more DNA than the other.

DNA puffs have been found too in the polytene chromosomes of foot
pad cells in dipterans. In *Sarcophaga bullata,* for example, there are two
giant cells per pad which secrete the cuticle of the adult fly during
pupation and DNA puffs have been reported on at least two of the
chromosomes in these cells. In addition, DNA granules are present at
specific times during the pupation phase. These too arise from the
polytene chromosomes by differential replication but in many cases the
regions from which they originate are unexpanded. The function of
these granules is not known but since a large quantity of material must
be produced and secreted by these two cells in a relatively short space of
time, there is the interesting possibility that they might serve to amplify
the mRNA production of specific loci in a manner reminiscent of the
amplification of rDNA.

Perhaps the most extensive system of amplification so far described
is that in *Hybosciara fragilis.* Here da Cunha, Pavan, Morgante, and
Ganido (1969) have shown that SG (salivary gland) nuclei contain large
numbers of micronucleoli in young larvae. They are produced at several
distinct regions of the chromosomes where they occur singly or in
groups. Single micronucleoli are associated with well-defined bands
while associations of micronucleoli in groups can be found in regions of
four different kinds:

    (a) at the nucleolar puff;

    (b) at the centromere region of the X-chromosome;

    (c) at the tip of an autosome which carries a duplication in the form
of two terminal bands; and

    (d) on a large inverted duplication involving $\frac{1}{4} - \frac{1}{5}$ (c. 40 bands) of
the length of one of the chromosomes. This duplication has the
appearance of a giant puff during most of larval life.

The central granules of the micronucleoli are Feulgen-positive and
[^{3}H] UdR autoradiographs indicate that RNA is also being made. These
micronucleoli are later released into the nucleoplasm.

Gall midges (Cecidomyidae) are distinguished by a regular differentiation in chromosome number between germ-line and soma. For
example, in *Wachtliella persicariae*, germ line cells contain 32 so-called
E-chromosomes in addition to the 8 standard or S-chromosomes which

are present in the soma. The E-chromosomes are regularly eliminated from the soma during early development. In oogonia the S- and the E-chromosomes are morphologically distinguishable. The S-chromosomes are heteropycnotic while the E-chromosomes are despiralized. Moreover at meiosis only the S-chromosomes pass through a conventional meiotic cycle. The E-chromosomes form four diffuse, cloud-like bodies.

At diplotene, concentric lamellae form around each S-bivalent separately but later they merge into common lamellar structures which surround the four S-bivalents (Fig. 27). During diakinesis each of the

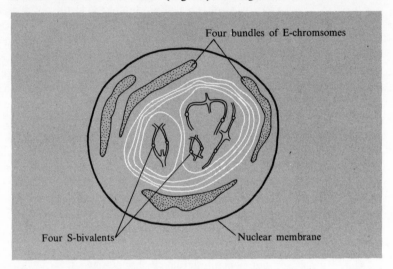

FIG. 27. Diplotene of female meiosis in *Wachtiella* (after Kunz, Trepte, and Bier (1970)).

four bundles of E-chromosomes contracts to form loose groups but this contraction is short-lived and they despiralize once again to the point of invisibility. This second despiralization affects the S-chromosomes too and they become increasingly diffuse until a point is reached where they cannot be distinguished from the E-chromosomes. The correspondence between the number of bundles of E-chromosomes and the haploid number of S-chromosomes is striking and suggests that the E-elements may have originated by polyploidization of the S-group. Finally the four bundles fall apart into single E-chromosomes which, together with the S-chromosomes, form 20 bivalents which condense and orient on a first division spindle.

When [^{3}H]UdR is applied to meiotic cells, the labelling is immediately restricted to the peripheral region where the bundles of E-chromosomes are situated. When the S-chromosomes despiralize, RNA synthesis can also be detected in the centre of the nucleus in the region of the S-chromosomes. Since electronmicrograph sections do not indicate the presence of nucleoli either in the region of RNA synthesis, or indeed elsewhere, Kunz, Trepte, and Bier (1970) have speculated it is mRNA that is being formed. But whatever its nature the indications are that an escalation mechanism is operative here too.

3.2. Inactivation mechanisms

Female mammals with a simplex X-system (see p. 41), and heterozygous for sex-linked coat-colour genes, commonly show a variegated or mosaic phenotype with patches of mutant as well as wild-type coat. Human females heterozygous for the locus determining glucose 6-phosphate dehydrogenase (G6PD) deficiency have two populations of erythrocytes, one with a normal enzyme level and the other with a very low level comparable to that of the recessive homozygote or male hemizygote (XY). This applies too in the case of the enzymes hypoxanthine guanine phosphoribosyltransferase (HGPRT) and α-galactosidase.

To account for this behaviour Lyon (1962) has proposed that:
(a) the simplex-X is epigenetically inactive when it is heteropycnotic (see p. 20) and, since one of the two X-chromosomes in normal females is in this state, both males and females, in effect, have only one functional X-chromosome in somatic cells;
(b) X-chromosome inactivation takes place at the time of implantation (= blastula stage) and is usually random so that in some cells the maternally-derived X is inactivated while in others the paternally-derived X is affected;
(c) this inactivation occurs only during a restricted stage of the developmental cycle but subsequent multiplication by mitosis leads to whole patches or populations of cells in the adult which show the same pattern of X-inactivation as the parental cell initially subjected to inactivation. On this argument heterochromatinization once determined is irreversible.

The hypothesis says nothing concerning the rates of multiplication of the two types of cell, the number of divisions they undergo, their migration, mingling, or interactions with other cells. Therefore, whereas the hypothesis does require that two types of cell should be present in the adult, it does not require that they should be in exactly equal numbers nor that they should be randomly distributed. Indeed, under certain circumstances the decision as to which X becomes inactive appears to be non-random. Thus when the female has one normal X and one abnormal X there is a preferential inactivation of the abnormal one, or at least, cells with a functional normal X-chromosome predominate in the adult.

Independent evidence of a physiological differentiation between the two X-chromosomes in female mammals of the simplex type comes from the finding that the two Xs have different patterns of DNA replication, the heterochromatic X replicating later and faster than the euchromatic

X-chromosome (see p. 40). The process of inactivation is believed to serve as a system of dosage compensation so that, although there is a difference between the sexes in the dosage of the X, there is only one active X in the adult somatic cells of both males and females.

Lyon's (1962) hypothesis thus implies that the production of the X-chromatin body, the delayed DNA synthesis of the condensed X, and its epigenetic inactivity are all related properties. Four lines of evidence can be used to support this statement.

(a) Cattanach (1961) succeeded in producing an X-autosome translocation by treating wild-type CBA mice with trimethylenemelamine (TEM). The translocation in question involved the transposition of a third of a medium-sized autosome representing linkage group I onto the X-chromosome (Fig. 28). The transposed section contained the dominant wild-type alleles of the recessive loci chinchilla (c^{ch}) and pink-eye dilution (p). The addition of the translocated section of the X led to a marked increase in the length of the X ($\approx$ 18 per cent). Consequently, while the normal X-chromsome X^N is no longer than

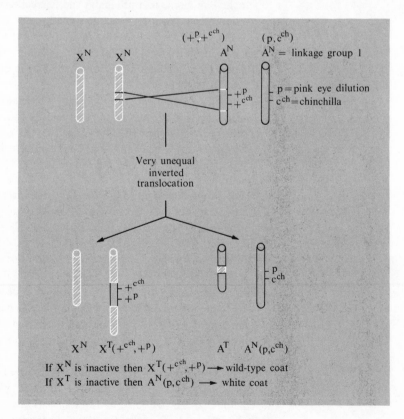

FIG. 28. Cattanach's X/A translocation in *Mus musculus.*

the third-largest autosome, the transposed X-chromosome X^T is larger than any other member of the complement and so is easily distinguishable from the X^N in somatic tissues.

Although the theory of X-inactivation was originally formulated on the basis of genetic data on the mouse, no sex chromatin can be demonstrated in this species because the sexual dimorphism of somatic interphase nuclei is obscured by the occurrence of several coarse chromocentres. Each autosome of the mouse carries a small block of heterochromatin adjacent to its centromere and these tend to aggregate as chromocentres which approximate to the expected size of the X-chromatin or Barr body. The condensed X, however, can be readily demonstrated in prophase figures of the female mouse.

Both $X^T X^N$ females and abnormal $X^T X^N Y$ males show a variegated phenotype of the category termed flecked with patches of wild-type (+) coat and patches of white (c^{ch}) coat. But neither $X^T Y$ males nor abnormal $X^T O$ females are variegated. A study of somatic prophases from the skin of flecked females of the type $X^T X^N$ shows that the heterochromatic X in the white patches is larger than that in the dark areas. That is, wild-type patches contain a condensed inactive X^N and an isopycnotic X^T while white patches are populated by cells with a condensed and inactive X^T and an isopycnotic X^N. Equivalent observations have also been made in $X^T X^N Y$ males where wild-type patches were shown to carry a heterochromatic X^N while in white patches the larger X^T chromosome was heterochromatic. By contrast in skin from wild-type $X^T Y$ males and $X^T O$ females there is no condensed chromsome. It would appear, therefore, that this constitutes a case of regional supression of the activity of autosomal genes when these are translocated onto the heterochromatic X of the mouse.

(b) Autoradiographs of female mice of the type $X^T X^N$ can be used to test whether late labelling with [^{3}H] TdR is random. In such an experiment Evans, Ford, Lyon, and Gray (1965) found 92 cells with a 'hot' X^T chromosome and 88 cells with a 'hot' X^N chromosome following [^{3}H] TdR uptake.

(c) There appears to be a general correlation, therefore, between positive heteropycnosity, delayed DNA synthesis, and the inability to take up [^{3}H] UdR. That is, condensed regions are unable to transcribe. This has been demonstrated at meiosis for the univalent X-chromosome of the male locust, *Schistorcerca gregaria,* and for the X-chromosome of male mice.

(d) The female mule has one X-chromosome derived paternally from the donkey and one derived maternally from the horse. Although the two are the same size, they are morphologically distinct in that the donkey X has a subterminally located centromere while the horse X is metacentric. Two kinds of fibroblast clones can be derived from each female mule. In one the donkey X is heterochromatic and late-replicating and this type produces only horse G6PD. In the other the horse X is late-replicating and only donkey G6PD is formed (Fig. 29).

The correlation between heterochromatinization and inactivation is supported by studies of two other kinds. First, in mealy bugs (*Hemiptera: Homoptera*) the entire set of paternally derived chromosomes becomes heterochromatinized in all male embryos at the blastual stage. During the subsequent development of the male, the paternal set retains its heterochromatic character up to and even during meiosis, it is regularly late labelling, and shows no sign of transcriptional activity. Throughout all the mitoses the heterochromatic set divides synchronously with the euchromatic set. At meiosis, however, there is no pairing between the maternal and paternal homologues at first prophase and all the chromosomes divide in a mitotic-like fashion at first anaphase. This is followed by a second division when the heterochromatic and euchromatic sets segregate from one another, although there is no pairing between them. Two of the four haploid nuclei so produced contain a chromosome

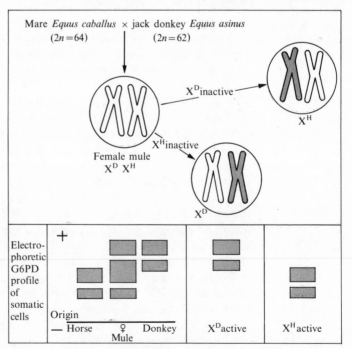

FIG. 29. Patterns of X-inactivation in the female mule. The glucose 6-phosphate dehydrogenase (G6PD) system is X-linked in both the horse and the donkey. Moreover the X-chromosomes differ morphologically; the donkey X (X^D) is submetacentric while the horse X (X^H) is metacentric.

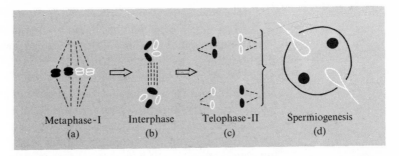

FIG. 30. Spermatogenesis in the mealy bug. Note there is no pairing at any stage of meiosis. The maternal (white) and paternal (black) sets both divide equationally at the first meiotic division and then segregate at the second division. Only the maternal products form sperm; and paternally derived spermatids degenerate (after Brown and Nur (1964)).

set derived exclusively from the mother while the other two have exclusively paternal chromosomes. (Fig. 30). Only the former proceed to form functional sperm.

In the armoured scale insects the process has been carried further and the paternal set is eliminated during one of the early cleavage mitoses (see p. 79). Notice that in both female mammals and coccid bugs, inactivation seems to occur at about the time that the embryo begins to synthesize its own mRNA.

A second line of study in relation to the role of heterochromatinization in the regulation of gene activity is that on V-type position effects. In *Drosophila*, it has been known for many years that various genes sometimes fail to function if they are transposed to certain unaccustomed locations by chromosome rearrangements. Consider, for example, a genic heterozygote +/m where the dominant wild-type gene (+) is carried on a rearranged chromosome while the mutant, allelic gene (m) is borne by a normal chromosome. In certain situations of this kind, the wild-type gene succeeds in functioning normally in only some of the cells. As a result a mosaic or variegated phenotype is produced in which some parts are wild-type while others are mutant. This constitutes the so-called V-type position effect and, like the mosaicism of female mammals, it is obvious, of course, only in relation to genes whose products are not transported between cells. V-type position effects are not found when the locations of the alleles are reversed, i.e. when the wild-type gene is on the normal chromosome and the mutant gene on the rearranged member. They do arise, however, in individuals which are homozygous for chromosome rearrangements so, clearly, the effect does not depend on structural heterozygosity.

Rearrangements lead to V-type position effects only when:
(a) one of the break points involved in the rearrangement occurs in the heterochromatic region; and

(b) the gene in question is brought into contact or, at least, close proximity with heterochromatin as a result of rearrangement.

The expression of over 70 different genes is known to be subject to positional influences leading to variegation in the phenotype. For example, when the wild-type allele for red eye (w^+) is placed near heterochromatin by rearrangement many of the eye facets have little or no pigment (w^v); white variegated phenotype. In consequence, w^+/w animals have red-white mosaic or variegated eyes when the wild-type gene is coupled with an appropriate rearrangement. Moreover, the suppressive effect of the heterochromatin can extend or spread an appreciable distance along the chromosome. Thus, in *D. melanogaster* such a spreading effect can involve a crossover distance of 6 per cent ($\equiv 50$ polytene bands) though loci further from the heterochromatic site are progressively affected both less and later in development.

V-type position effects are reversible, in that normal gene function is restored when the gene is removed from the proximity of heterochromatin. They are also highly sensitive to changes in the content and arrangement of heterochromatin elsewhere in the nucleus. Therefore, the effect must be on the expression of the gene and not its organization. Finally, genes normally located in or near heterochromatin (e.g. $1t^m$, light mottled, in *D. melanogaster*) appear to be dependent on such proximity for their 'normal' expression in that they display variegation when transposed to euchromatic regions.

Position effects show how the function of a gene can be influenced by the genetic constitution of other regions of the same chromosome especially those in the immediate vicinity. They substantiate the view that the genes of a chromosome do not function independently and demonstrate that the precise linear continuity of genes is important in determining the pattern of gene action. In other words, there is a measure of co-ordinate gene function. However, numerous cases of translocation of genes from their normal position are known which do not seem to result in an alteration of genic expression. This suggests that either there is a class of genes which are autonomous in function such that they can exert their effect independently of their position or else that only rearrangements which do not lead to position effects are likely to prove of evolutionary significance. Nevertheless, where they occur, V-type effects can be interpreted in terms of the influence of heterochromatin on gene expression, that is, in terms of chromosomal regulation of gene function.

As early as 1947, Prokofyeva-Belgovskaya had in fact claimed to have seen a heterochromatinization of euchromatic bands adjacent to heterochromatin in a series of mottled position effects, so that instead of finding clear salivary bands, she observed a diffuse chromatic network. At that time, however, the suggestion that the function of a euchromatic locus could be impeded by heterochromatinization was regarded as crude and naive.

3.3. Elimination mechanisms

We have already commented on the fact that there is an under-replication of heterochromatic regions in polytene nuclei (see p. 69). A similar phenomenon occurs also in coccid bugs. Thus, in *Planococcus,* most tissues of the male contain 5 maternal euchromatic and 5 paternal heterochromatic chromosomes. The endopolyploid oenocytes and testis sheath cells also contain five heterochromatic chromosomes but, in addition, have up to 80 euchromatic ones, a clear indication that while the maternal euchromatic elements replicate in endomitotic cycles, the heterochromatic paternal ones do not. Similarly in the plant *Rhinanthus,* the chromosomes of the standard complement form a system of polytene units in haustorial cells of the endosperm but the small supernumerary or B-chromosomes do not.

More striking than these cases of under-replication, however, are those which involve an actual elimination of chromosomes. In some coccid species, as we have already mentioned (see p. 77), the entire paternal set is eliminated during one of the early cleavage division cycles. More commonly, elimination is differential in that it occurs in the soma but not in the germ line. Two main variants are known, the one involving a differential loss of chromosome segments and the other the differential loss of entire chromosomes.

3.3.1. *Loss of heterochromatic segments*

In the horse threadworm, *Parascaris equorum,* the gametic chromosomes, unlike those of somatic cells, are compound polycentric units which are terminated at each end by long acentric and heterochromatic extremities. At the first cleavage mitosis, these compound entities are distributed normally, but from the second to the fifth cleavage division there is a series of elimination mitoses in which the heterochromatic ends are detached and the compound polycentric euchromatic sections dissociate to give a large number of monocentric elements. By the 16-cell stage, only one of the cells retains the compound chromosomes intact and this represents the anlage of the definitive germ line (Fig. 31).

Painter (1966) has suggested that the retention of the heterochromatic segments in the germ line is related to the production of a population of ribosomes during female gametogenesis to provide for the needs of early embryonic development. In the soma, where such a population is not required, the heterochromatic segments can be dispensed with. Kaulenas and Fairbairn (1966), on the other hand, have shown that male and female pronuclei of *Parascaris* remain unfused during the first 50–60 hours of development, and that in the 12–24 hour period large amounts of rRNA accumulate around the male pronucleus. They argue, therefore, that the sperm nucleus produces rRNA after activation of the egg so

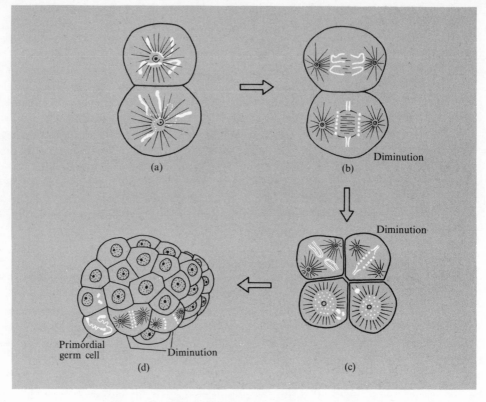

FIG. 31. Chromosome diminution and the formation of the germ line in
Parascaris equorum (after Boveri (1892)).
  (a) Second cleavage metaphase;
  (b) second cleavage anaphase;
  (c) third cleavage metaphase;
  (d) thirty-two cell stage, only the primordial germ cell now carries the compound
  chromosomes.

that it begins cleavage with stored rRNA of paternal origin.

The only other sensible explanation for the differential retention of
heterochromatin in the germ line is that this material is in some way
concerned with the differentiation of the germ line itself. The germ-cell
lineage in animals can usually be traced to the pregastrula period of
development. Indeed, in some cases, as in *Parascaris,* the germ line is
defined during the first few cleavage divisions (Table 11). In all these
cases, the germ lineage appears to be defined according to the
localization of an egg cytoplasm determinant and, of particular importance, the cytoplasm of the germ-cell anlage contains a large amount of
nucleic acid. Wallace, Murray, and Langridge (1971) have, therefore,
suggested an alternative explanation for gene amplification in amphibians,
based on this situation. The theory of gene amplification, you will recall

(see p. 67), asserts that the rRNA of the oocyte nucleolus organizer
undergoes extra replication to yield some 1500—4000 copies of the
organizer, in anticipation of the intense rRNA and protein synthesis
which is subsequently required. But as Wallace *et al.* (1971) point out,
both oogonia and spermatogonia of the young toads also contain
supernumerary nucleoli and excess rRNA. The extra nucleoli of
spermatogonia appear to be identical to those of oogonia but lack their
postulated justification. Wallace *et al.* have, therefore, argued that the
excess DNAs are self-replicating independent satellites restricted to the
germ line and containing reiterated rDNA sequences.

Table 11

Differentiation of the germ-cell line in animals

Organism	Appearance of germ-cell anlage
Frog	Prior to 8-cell stage
Ascaris Insects	16-cell stage
Molluscs	Pre-gastrula

Each germ-line cell thus contains a population of independent rDNA
satellites which are not attached to chromosomes and are not strictly
copies of the organizer except perhaps in evolutionary terms. When in
the nucleus, these satellites can replicate and transcribe RNA but at the
first meiotic division they are banished into the cytoplasm where they
neither replicate nor transcribe. In the male they are then presumably
discarded, with the excess of cytoplasm, during sperm maturation, but
in eggs they collect as a constituent of the germinal plasm at the vegetal
pole soon after fertilization and then pass into certain embryonic cells
where they serve as germ-cell determinants. Once the germ line is defined,
these satellites re-enter the nucleus and resume their function of producing
rRNA when mitosis starts again in oogonia and spermatogonia.

In the Wallace theory, therefore, the rRNA particles behave like
episomes which persist in the germ line. The principal weakness with this
argument is that rDNA can be shown to be amplified by the chromosome
in the oocyte stage though this does not preclude both amplification in
the conventional sense and persistence, as proposed by Wallace.

Beerman (1966) has described a quite different kind of elimination
process in the copepod genus *Cyclops* which again leads to differences
in the heterochromatin of germ line and soma. In *Cyclops strenuus,*
for example, all the diplotene chromosomes of the oocyte have long
terminal segments of heterochromatin which first become distinguishable
at the telophase of the fourth cleavage mitosis. During the fifth cleavage
division, these segments are expelled into the spindle, as a result of

which the chromosomes suffer a drastic reduction in length. This change is irreversible so that no heterochromatin is found in somatic cells after the fifth cleavage. The chromosomes in the primordial germ cells and their descendents remain unchanged, however.

In *Cyclops furcifer,* the chromosomes have heterochromatic segments both near the central centromeres and at the ends of the chromosomes: the heterochromatin in both regions is subject to diminution. Since the chromosomes remain metacentric and do not appear to be broken by the diminution process, it is clear that interstitial material can be removed from the chromosome without disruption of linear integrity (cf. nucleolar rings of plethodonts). The extent of loss in these cases of diminution is considerable, for there is a 65 per cent difference in DNA content between post- and pre-diminution stages in somatic cells.

A distinctive form of diminution occurs during the formation of the macronucleus in the hypotrichous ciliate *Stylonchia*. The first stage in the formation of the macronucleus from a micronucleus involves the formation of polytene chromosomes in the macronuclear anlage. These polytene chromosomes then break up into short lengths by interband breakage, each band becoming enclosed in a separate vesicle. Immediately after the break-up process some 93 per cent of the DNA in the anlage is eliminated. The remaining 7 per cent then undergoes many rounds of replication until finally the DNA content of the macronucleus is increased to about 65 times that of the micronucleus. By comparing bouyant densities and thermal denaturation curves, Bostock and Prescott (1972) have estimated that more than 60 per cent of the DNA sequences in the micronucleus are missing from the definitive macronucleus.

3.3.2. *Loss of entire chromosomes*

In the creeping vole, *Microtus oregoni,* male zygotes are $X^M Y^P$ while female zygotes are $X^M O$ in constitution, M and P serving to distinguish maternal and paternal derivation. By selective non-disjunction in the germ line, the female becomes $X^M X^M$ so that all eggs transmit an X^M. The female soma remains $X^M O$. The X^M, on the other hand, is eliminated from the male germ line which thereby becomes YO in constitution. Here then the female has dispensed completely with one of the X-chromosomes in the soma which is not just functionally XO as in most female mammals but genetically XO as well.

More common than cases of complementary non-disjunction are systems involving complementary gametic elimination (Table 12). This mechanism has been used to stabilize uneven polyploids or complex chromosome heterozygotes and we shall return to consider examples of it when we deal with chromosomes from an evolutionary aspect (see p. 116).

Table 12

Some examples of complementary gametic elimination in plants and animals

Species	Soma	Gametes ♀	Gametes ♂
Plants			
Rosa canina (2n = 5x = 35)	AABCD (35)	ABCD (28)	A (7)
Leucopogon juniperinum (2n = 3x = 12)	AAB (12)	AB (8)	A (4)
Oenothera muricata	aα	a	α
Animals			
Enchytraeus lacteus (2n = 170)	162S + 8L	162S + 4L	4L
Sciara coprophila (2n = 6A + 2Xp + 1Xm)	♂6A + Xm ♀6A + Xp + Xm	3A + Xm	3A + 2Xp
Microtus oregoni (2n = 16 + XY♂ 16 + XO♀)	♂16A + XY ♀16A + XO	8A + X	8A + Y 8A

Note. Detailed explanations of the constitution of soma can be found in John and Lewis (1968).

3.4. Cell differentiation

The many cells of a multicellular eucaryote are differentiated structurally and functionally into various distinct types. For example, at least a hundred different cell types can be distinguished among the 10^{13} cells which make up the human body. These types differ in shape and size, and form and function and their distinction is seen most clearly in the diversity of their biochemical pathways (Table 13).

Although the meristematic cells of plants and the stem cells of animals retain the zygotic chromosome complement, this is not always true of cells that do not undergo further mitosis. Such cells may grow by the much simpler process of endomitosis involving either polyteny or endopolyploidy. In the former case the products of repeated replication remain in association (see p. 35) while in the latter case the products separate from one another. Even so the nuclei of many of the cell types found in eucaryotes are direct mitotic descendents of the zygote nucleus. Each, therefore, must contain the same genetic information, for the mitotic mechanism is designed to achieve an exact qualitative and quantitative distribution of the products of chromosome duplication. This, in turn, must mean that differentiated cells utilize only a small fraction of their total genetic information and that the process of differentiation must be concerned with the regulation of the genetic potentialities of the cell. In biochemical terms this implies that only

Table 13

Biochemical diversity of human cell types

	Cell type	Specific biosynthesis
A. Ectodermal derivatives	1. Epidermis	Keratin, cholesterol
	2. Adenohypophysis	
	(a) Eosinophils	Growth hormone, lactogenic hormone
	(b) Basophils	Thyrotrophic hormone, ACTH, luteinizing hormone, FSH
	3. Neurons	Complex lipids, acetylcholine
	4. Modified nerve cells	
	(a) Andrenal medulla	Norepinephrine, epinephrine
	(b) Intestinal argentaffin	Hydroxytryptamine
	(c) Melanocytes	Melanin
	(d) Neurohypophysis	Vasopressin, oxytocin
B. Endodermal derivatives	1. Salivary gland	Ptyalin
	2. Peptic cells	Pepsinogen
	3. Pancreas	
	(a) Exocrine	Endopeptidases, nucleases
	(b) Endocrine	Glucagon, insulin
	4. Intestinal mucosa	Exopeptidases
	5. Liver	Bile acids
	6. Thyroid	Thyroxine, thyroid hormone
C. Mesodermal derivatives	1. Muscle	Myosin
	2. Erythrocytes	Haemoglobin
	3. Adrenal cortex 4. Gonads	Steroid hormones
	5. Connective tissue	
	(a) Cartilage	Mucoproteins
	(b) Fibrocytes	Collagen, elastin

particular proteins are synthesized in particular cells at particular times. For example, the enzyme lactate dehydrogenase (LDH) enables cells to function effectively during transient deficiencies in oxygen availability for it converts pyruvic acid to lactic acid. When oxygen is again available the lactic acid can be reoxidized to pyruvic acid which can be further oxidized to CO_2 and H_2O.

LDH is a polymeric enzyme, a heteromultimeric tetramer of subunits each with a molecular weight of c. 35 000. It occurs in multiple molecular forms, or isozymes, not only in different individuals but even in different tissues of the same individual. The multiple forms within individuals represent different combinations of two basic subunits, termed A and B, which are synthesized by two distinct genes. Five

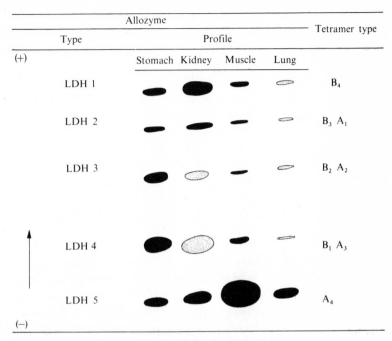

Allozyme						Tetramer type
Type	Profile					
(+)		Stomach	Kidney	Muscle	Lung	
LDH 1						B_4
LDH 2						$B_3 A_1$
LDH 3						$B_2 A_2$
LDH 4						$B_1 A_3$
LDH 5						A_4
(−)						

FIG. 32. Electrophoretic profiles of LDH allozymes in the rat.

tetramers are commonly recognized and referred to as LDH1−5 respectively, the tetrameric constitutions being B_4, $A_1 B_3$, $A_2 B_2$, $A_3 B_1$, and A_4. When separated on a chromatograph profile or zymogram these occur in a specific sequence (Fig. 32).

The eggs of mice are initially equipped only with LDH1, the B_4 tetramer, which implies that only the B-gene is active in oogensis. As development proceeds, the action of the B-gene is suppressed and the A-gene is activated so that the isozyme pattern is skewed towards the LDH5 end of the spectrum. During late embryonic development and especially with the approach of birth, the B-gene is progressively reactivated in many tissues so that the isozyme profile again shifts towards the LDH1 end of the spectrum. These changing patterns reflect shifts in the relative activities of the A- and B-genes. Moreover, the ratio of isozymes is characteristic for each type of tissue suggesting that the A- and B-subunits are synthesized in different amounts in different tissues.

Spermatogonia are equipped with both A- and B-subunits, but the activity of both the A- and the B-genes is suppressed at the inception of spermatocyte differentiation. At this stage a third locus, the C-gene, is activated to produce a C-tetramer with quite distinct properties. This gene is inactive in all cells of the body except primary spermatocytes and, even in these cells, it is active for only a brief period. This C-gene has been found in most mammals and birds while fish have an E-gene which is turned on at only a few specific stages in development and only in a few kinds of cells, namely the retina and certain nerve cells of the brain.

An equivalent picture emerges for haemoglobin in man. Here at least 4 genes, α, β, γ, and δ, encode the subunits of human haemoglobin while a fifth, the ϵ-gene, contributes to its synthesis in the embryo. These genes again function at specific stages and their relative activity determines the relative abundance of the different tetrameric forms of haemoglobin.

3.4.1. Repression

The concept of the operon has served to emphasize repression as one of the main mechanisms of gene control in procaryotes. It is now over 20 years since Stedman and Stedman first postulated that histones might serve as repressors of gene activity in eucaryotes. Although at first dismissed as speculative, their hypothesis appears to contain the essence of one mode of gene regulation.

Chromatin can be isolated from resting nuclei in a form which can sustain the synthesis of RNA when the necessary precursors, enzymes and energy are provided exogenously. What is more, the RNA so generated is capable of supporting the synthesis of protein in the presence of ribosome and other necessary components. Chromatin isolated from differentiated cells is, however, far less active in supporting RNA transcription than pure DNA prepared from such chromatin (Table 14). Evidently, therefore, the DNA of chromatin from differentiated cells is in part repressed and not available for transcription.

Table 14

The relative templating activity of liver chromatin fractions

Template	Chemical composition		RNA synthesis (AMP incorporated pm/0·5 ml/10 min)
	Histone/DNA	Non-histone protein/DNA	
1. Isolated chromatin	0·94	0·64	2580
2. Purified DNA prepared from isolated chromatin	0·00	0·02	9100
3. Isolate chromatin treated with 0·2N HCl*	0·03	0·55	7800

* Acid extraction results in selective removal of histone, but not of non-histone, protein.
Note. Data from Marushige and Bonner (1966)

That the histone component of chromatin plays a role in determining this lowered template activity is demonstrated by two facts:

(a) the percentage templating activity in nuclei with different histone contents is inversely proportional to the amount of histone present (Table 15); and

Table 15

The templating activity of chromatin from different sources

Chromatin source	Content relative to DNA of:			Templating activity as % of DNA
	Histone	Non-histone protein	RNA	
1. Pea vegetative bud	1·30	0·10	0·11	6
2. Rat ascites tumor	1·16	1·00	0·13	10
3. Cow thymus	1·14	0·33	0·007	15
4. Sea urchin blastula	1·04	0·48	0·039	10
5. Pea embryonic axis	1·03	0·29	0·26	12
6. Human HeLa cells	1·02	0·71	0·09	10
7. Rat liver	1·00	0·67	0·043	20
8. Sea urchin pluteus	0·86	1·04	0·078	20
9. Pea growing cotyledon	0·76	0·36	0·13	32

Note. Data from Bonner, Dahms, Farnbrough, Huang, Marushige, and Tuan (1968).

(b) selective removal of histone, unlike that of non-histone protein, increases the template activity of chromatin to that of de-proteinized DNA. It would appear, therefore, that gene activity can be repressed by the presence of histone and correspondingly de-repressed by its removal.

Detailed analysis of acid-extracted histone from calf thymus lymphocyte nuclei shows that histone is in fact a mixture of four principal classes which are distinguishable in amino acid composition (Table 16). By centrifuging nucleoprotein through salt solutions of appropriate concentrations, particular classes of histone can be selectively removed from calf-thymus lymphocyte nuclei. It is also possible to reconstitute nucleohistone by combining a particular histone fraction with purified DNA. Such reconstituted nucleohistones show a gradation in the effectiveness with which they support RNA synthesis (Table 17). Different histone fractions thus differentially affect the DNA profile of chromatin; in particular, classes Ib and IIb appear to be especially effective in repression.

Much of the chromatin in the nuclei of calf thymus lymphocytes is arranged in dense heterochromatic masses. If these nuclei are isolated and incubated with [³H] UdR, label is incorporated only by the diffuse chromatin. This implies that gene activity is repressed in the condensed regions and that something like 80 per cent of the total chromatin must be in a repressed state.

Both condensed and diffuse chromatin have the same histone composition, with some 20 per cent of the total comprising the lysine-rich

Table 16
Amino-acid composition of calf thymus histone fractions

| Amino acid | Histone fraction | | | | Whole thymus histone |
| | Lysine rich | Moderately lysine-rich | Arginine-rich | | |
	Ib	IIb	III	IV	
Lys	26·2	13·5	9·4	9·0	16·0
His	0·2	2·8	1·6	1·6	1·5
Arg	2·2	7·9	12·9	12·8	7·9
Asp	2·5	5·6	4·4	4·5	4·7
Thr	5·4	5·2	7·4	7·4	5·5
Ser	6·5	7·0	4·1	4·1	5·2
Glu	4·3	8·7	9·9	10·6	8·0
Pro	10·2	4·7	3·8	4·2	5·9
Gly	7·3	8·2	8·8	7·9	8·3
Ala	24·5	11·5	11·8	12·3	15·1
Cys	0·0	0·0	0·0	0·0	0·0
Val	4·1	6·7	5·8	5·6	7·0
Met	0·1	0·8	1·2	1·2	0·7
Ile	1·2	4·5	5·4	5·4	3·5
Leu	5·0	8·6	8·7	8·9	7·6
Asn	0·6	1·3	2·5	2·7	1·4
Tyr	0·7	3·0	2·4	2·3	2·2
Try	0·0	0·0	0·0	0·0	0·0
Basic AA/ acidic AA	4·9	1·9	1·5	3·9	

n.b. Lysine-rich histone Ib binds weakest to DNA, arginine-rich histone binds strongest.

Ib fraction. When this lysine-rich fraction is selectively removed from isolated nuclei, the condensed chromatin dissociates into a diffuse network of fibrils while the removal of the arginine-rich histone fraction has no such effect. Clearly, the relationship between the lysine-rich histone and DNA in condensed chromatin is quite different from that in diffuse chromatin, but the nature of this difference has not been

Table 17
Relative activity of reconstituted nucleohistones in support of RNA synthesis by calf thymus extracted material

Template	RNA synthesized (pm nucleotide/0·5 ml/10 min)
DNA alone	8474
Nucleohistone Ib	56
IIb	140
IV	4000

Note. Data from Huang (1964).

clarified. Selective methods for removing all but arginine-rich histones
from calf-thymus chromatin reveal that they are bound to regions very
rich in GC base pairs. Histone proteins extracted from calf-thymocyte
nuclei are also capable of repressing the template function of DNA in a
variety of cell-free systems too (see p. 63), and equivalent effects are
known in other organisms (Table 18).

Table 18

*Relative activities of apical bud chromatin fractions of pea plants in
support of globulin synthesis*

Template	Pea seed globulin as % of total soluble protein formed
1. Native apical bud chromatin	0·13†
2. Apical bud chromatin after removal of histones I and IIb	1·73
3. Apical bud chromatin after removal of all histones	0·38

Note. Data from Bonner and Huang (1966).
† n.b. The immunochemical assay used to detect globulin has a background
detection level of 0·13.

Histones, however, are a remarkably uniform group of molecules.
Not only are the histones of different cell types within an organism very
similar to one another in composition, but so are those of different species.
The differences that do exist at the specific level appear to involve only
single amino acid residues just as do those differences between corresponding enzymes, haemoglobins, and insulins in different species. Thus
calf and pea histone IV molecules differ by only two amino acids and
even these are conservative substitutions. It follows then that the same
histones must be capable of repressing different genes in different cell
types. It is true that in several totally repressed systems like nucleated
erythrocytes and sea urchin sperm, histone I is replaced by histone V.
Amino acid analyses of corresponding preparations from fish, amphibian,
and avian erythrocytes show significant variation in this fraction V
particularly in lysine, serine, and arginine contents but otherwise histones
appear to lack the specificity one might have expected of a discriminating
repressor molecule.

From the known composition of DNA histone it can be calculated
that in a nucleoprotein in which DNA is fully complexed with basic
protein, the mass ratio of histone to DNA should approximate to 1·35:1.
It is clear, however, that chromatin, in general, does not contain enough
histone to complex fully with all the DNA (Table 15). Thus differences
in histone /DNA ratio may reflect differences in the extent to which the
genome is repressed in different tissues. Indeed, different regions of the

genome could be repressed in different cell types with the same histone/ DNA ratio according to the pattern of histone distribution along the DNA of the chromosomes.

The non-histone, acidic protein component of chromatin does not in itself appear to inhibit transcription. But it may exert a directing role by determining which histone-associated regions of the template are to be repressed. Experiments by Paul, Gilmor, Thomson, Threlfall, and Kohl (1970) suggest that histones have a relatively non-specific repressing effect on transcription. They also show that the non-histone proteins include molecules which can both antagonize the repressive effects of histone and independently repress transcription. Indeed they find that if two chromatins are reconstituted from DNA and histones of common origin, but with non-histone proteins of different origin, then the transcription product resembles that of the non-histone protein source, which implies that the transcriptional specificity of reconstituted chromatin is determined by the non-histone fraction.

The general picture that emerges therefore is that histones perform a non-specific inhibitory role and that where DNA and histones are associated no transcription occurs in the absence of other components. For example, during the conversion of a spermatid into a sperm there is a marked reduction in the amount of cytoplasm and a reorganization of the nucleus involving the loss of all acidic protein and RNA from it. In addition, in many species, a new basic protein (protamine) is synthesized which is rich in arginine. The integrity of the chromosomes is, of course, maintained during these changes but their organization is considerably simplified and they become synthetically silent. Thus, unlike the differentiation of the egg which is essentially a process of elaboration and expansion, that of the sperm is concerned with simplification and reduction.

Non-histone components may affect the binding of histones to DNA so as to permit the attachment of RNA polymerase for transcription. Finally, non-histone components which themselves mask DNA may contain factors which act as specific repressors. It may be significant that one of the first alterations during the onset of puffing in polytene chromosome bands (see p. 35) is the accumulation of a non-histone protein and this accumulation precedes the onset of transcription.

Bonner and Huang (1966) have described a special kind of RNA, chromosomal RNA, which appears to be covalently linked to protein, and they speculate that it might provide a recognition mechanism by base pairing with DNA. That is RNA may mediate in the sequence specific association of DNA and protein on the basis of DNA-RNA sequence recognition. Such specific associations might then either serve to bind histone to DNA and so repress transcription, or else prevent histones from binding to particular DNA sequences and so keep that sequence open for transcription.

The role of histone in repression and its correlated connection with condensed chromatin clarifies the long-established observation that during gastrulation, differences arise in the gross appearance of the resting nuclei in many tissues according to the pattern of specific heterochromatinization. Diploid mammalian liver nuclei, for example, contain large single heterochromatic granules. Nuclei of the renal cortex, by contrast, include a large number of small granules, while thymocyte nuclei show massive blocks of heterochromatin, and so on (see Fig. 5). This suggests that a specific pattern of repression is developed at gastrulation which is determined by, and reflected in, specific patterns of heterochromatinization.

In the systems we have covered so far, heterochromatinization is stable and once established is not reversed. In at least one case, however, a de-heterochromatinization has been observed. In coccid bugs the paternal set of homologues assumes a positively heteropycnotic state at the blastula stage. As development proceeds this process is reversed in several tissues and, as Nur (1966) has shown, the time of reversal is specific for the tissues concerned.

3.4.2. Activation

Differentiation cannot be explained solely in terms of selective repression of large sections of the genome. Coupled with this there must be specific patterns of activation in those sections which are unrepressed. The limited loci which remain available for activation in a given cell type include, of course, those genes which govern indispensable enzymatic machinery and which are presumably active in most, if not all, cells. But they include also those genes which are responsible for the specific enzymic repertoires whose collective activites define the metabolic patterns of particular differentiated states.

The most direct evidence for the activation of transcription in eucaryotes comes from work on the giant polytene chromosomes of dipteran flies. It is well established that puffs in these polytene units are sites of active RNA synthesis (see p. 38). In salivary glands of *Drosophila melanogaster,* a particularly intense period of specific puffing precedes puparium formation, more and larger puffs appearing in the polytene system. In *Camptochironomus tentans,* on the other hand, there is a shift only in the pattern of puff activity, the over-all number and size of the puffs remaining approximately constant. In both cases, the number of puffs that disappear entirely or else appear *de novo* is small. For example, in a study of 143 puffs in the salivary glands of *D. hydei,* Berendes (1968) found only 9 which disappeared or appeared anew. But 110 of them changed in size during either puparium formation or the subsequent pupal moult. In *Ch. tentans* about 10 per cent of the 1900 bands which can be distinguished seem to be in a puffed state at some stage during the last larval instar.

It can be assumed that those puffs which are present at all times, which change their activity little if at all during metamorphosis, and that are shared by many or all tissues, are involved in cellular functions common to all cells. These form by far the larger group of puffs. By contrast the smaller group of puffs that appear or disappear in preparation for the moult may mediate in reactions characteristic for a given step in moulting or metamorphosis. Steffenson and Wimber (1971) have used *in situ* molecular hybridization studies between [^{3}H] tRNA and the DNA in cytological preparations of salivary gland chromosomes of *Drosophila melanogaster* to locate genes coding for tRNA in about half of the total genome. They found that a number of the major sites correspond with chromosome segments which puff during larval or prepupal development.

Puffs can be induced experimentally by injection of the moulting hormone ecdysone, which is a steroid. Steroid hormones of vertebrate origin also appear specifically to alter gene transcription. The mode of action of these steroids, however, is not known with certainty. Kroeger (1964) has suggested that the primary action of ecdysone involves an alteration in membrane permeability which results in a change in the cellular Na^+/K^+ balance. This view is supported by a recent study on the influence of ionic strength and pH on isolated nuclei and chromosomes of salivary gland cells of *Ch. thummi*. All ions tested induced decondensation of bands, and this decondensation was correlated with an increase in template activity as evidenced by the incorporation of [^{3}H] UTP. Alternatively, the small non-polar steroid molecules presumably can pass fairly easily through the plasma membrane. It has been argued, therefore, that following their entry into the cytoplasm, steroid molecules may associate non-covalently with specific receptor or binding proteins to form complexes which then move into the nucleus where they may serve as specific activators of particular genes.

Endocrine control presumably also plays an important part in controlling the activity of the Y-chromosome in *Drosophila*. Each spermatid produced by meiosis has a relatively small nucleus which is inactive in replication and transcription. Nevertheless, the spermatid is transformed into a sperm and this transition is in reality a complicated kind of cell differentiation involving a great deal of synthesis and growth (Fig. 33). All these events take place under the control of preformed and stablized messenger molecues produced during the diploid phase. At least some of these messengers, in turn, arise from the activity of one chromosome—the Y. This shows a relatively short-lived phase of high synthetic activity involving five distinct gene loci. These loci enter into an active state during the pre-meiotic interphase when they produce loop-like structures which produce and accumulate mRNA (Fig. 34).

In somatic nuclei, by contrast, the Y-chromosome has no known activity. Indeed it can be dispensed with in somatic development, for

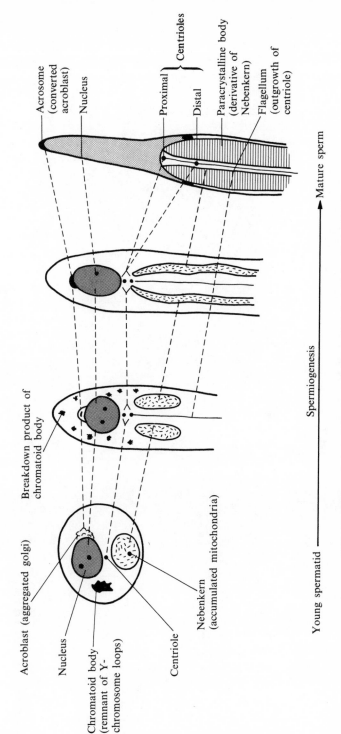

Acrosome (converted acroblast)

Nucleus

Proximal ⎱ Centrioles
Distal ⎰

Paracrystalline body (derivative of Nebenkern)

Flagellum (outgrowth of centriole)

Breakdown product of chromatoid body

Acroblast (aggregated golgi)

Nucleus

Chromatoid body (remnant of Y-chromosome loops)

Centriole

Nebenkern (accumulated mitochondria)

Spermiogenesis

Young spermatid ⟶   ⟶ Mature sperm

FIG. 33. Differentiation of the sperm in *Drosophila hydei* (after Hess (1967)).

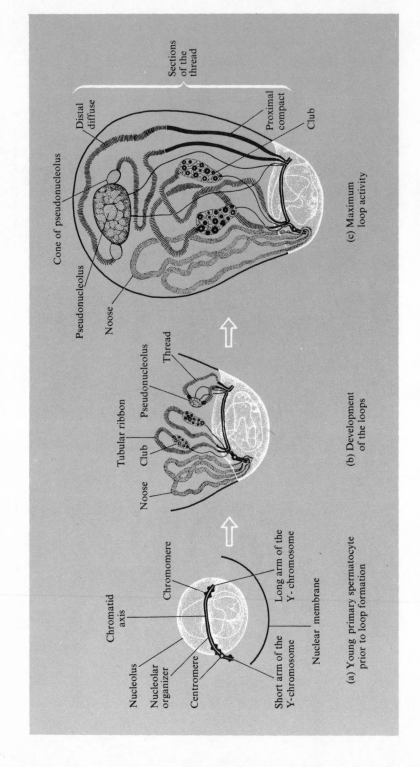

Nucleolus

Nucleolar organizer

Centromere

Short arm of the Y-chromosome

Nuclear membrane

Chromatid axis

Chromomere

Long arm of the Y-chromosome

(a) Young primary spermatocyte prior to loop formation

Noose

Tubular ribbon

Club

Pseudonucleolus

Thread

(b) Development of the loops

Cone of pseudonucleolus

Distal diffuse

Pseudonucleolus

Noose

Proximal compact

Club

Sections of the thread

(c) Maximum loop activity

FIG. 34. Lampbrush loop formation in the Y-chromosome of *Drosophila hydei* (after Hess (1971)).

XO males of *D. melanogaster* are fully viable. They are, however, sterile since, in the absence of the Y, normal sperm development cannot occur. Because of the lack of somatic activity, the Y-chromosome of *Drosophila* was, until recently, assumed to be inert. We can now see that, in reality, the Y-chromosomes includes genes which function exclusively in the germ line and specifically at the onset of meiosis. These genes are concerned solely with the control of sperm growth and organization. They must, of course, function well before their ultimate products are needed because only half the products of male meiosis will actually carry a Y-chromosome.

Clearly, therefore, the differentiation of one cell can be influenced by that of others which preceded it in its cell lineage and a gene can show an influence in a cell from which it has been removed by prior segregation or elimination.

4 *Phylogenetic functions*

**4.1. The components
of evolution**

Every breeding unit has the potentiality for producing many more
individuals of the same kind than can ever hope to survive. In most
cases the level of mortality is very high and survival conveniently can be
considered to have two principal components.

On the one hand, if genetically uniform individuals are subjected to a
nominally uniform environment which is incapable of supporting the
entire population, some mortality is inevitable. But, clearly, since the
survivors and the non-survivors cannot be different genetically, the
final population cannot be different from the initial one and the deaths
cannot be selective. Many of the deaths which occur under conditions of
gross overcrowding must be of this kind. This situation is closely approached
in those plants with closed breeding systems where the number of mature
seeds per ovary regularly falls far short of the number of ovules which are
produced and fertilized.

On the other hand, where individuals are sufficiently close together to
interact, a competitive element is almost inevitable and this will have a
genetic component in most cases. Further, individuals which are grossly
ill-adapted to a particular environment are not expected to survive even
in the absence of competition from phenotypes which are better suited
to the conditions in question. Survival therefore has a selective element
also, and the over-all level of mortality will depend on the interaction
between the selective and non-selective components.

But survival is only one aspect of fitness, for death in one generation
is but a measure of the effective fertility of the generation that preceded
it. Fitness, the contribution made to future generations, is thus dependent on both viability and fecundity. There is then selection by the
environment—via differential viability, differential fertility, or differential fecundity—of those characteristics which best fit the organism
to the conditions under which it lives. Selection is often concerned
with eliminating those chance changes in heredity which arise by
mutation, for these occur without reference to the adaptive or
adaptational needs of the organism. Less often it serves to preserve

and exploit such chance changes. The latter event promotes heritable adaptive change in relation to changing circumstances, while the former serves to maintain states of adaptation in an environment which is stable.

It is a truism, therefore, that living things are adapted to their environment—if they were not they could not survive. Adaptation, like fitness, must be regarded at two levels. First, it refers to the adjustments which occur within individual organisms as they respond through negative feedback to environmental changes during their own life-time—this is homeostasis or physiological adaptation. Secondly, adaptation refers to the temporal process of positive feedback in which transgeneration changes are directed by the selective pressure of the environment. This is evolution or genetic adaptation.

Those individuals which are similarly adapted in a genetic sense are inclined to be clustered in the same kind of environment, and the term 'population' is frequently used to describe such clusters of similarly adapted individuals. The term may, however, be further limited by restricting it to groups of sexual forms which associate for reproductive as well as ecological reasons. As genetically defined a population is a spatio-temporal group of conspecific interbreeding individuals. Such a group is endowed with a spatial integrity because its members interbreed. It maintains a temporal continuity because of the reproductive connection which exists between generations.

While all hereditary changes originate within the individual cells of individual organisms, these are not the units in which evolutionary change is ultimately observed. Rather evolution involves changes in the genetic structure of populations, and mutation and recombination are the principal means of producing the genotypes responsible for the phenotypes favoured by a particular environment. Although its level can be regulated within limits, mutation is an irrepressible property of genetic material of all organisms and, ultimately, it forms the basis of all inherited novelty in biological systems. It is augmented by the recombination which occurs when reproductive cells are formed or fused.

Recombinational processes are unknown in many prokaryotes though some have a wide variety of assortive mechanisms. These have been widely studied under experimental conditions but the extent to which they are utilized in nature is difficult to assess.

Genetic recombination in eucaryotes can be achieved at different levels at each of which some adjustment is possible (Table 19). Nevertheless, the over-all level can be selected in relation to specific requirements because many of the variables are under genotypic control. Those which apply during the production of the haploid phase of the sexual cycle affect the range of gametic novelty, while those which relate to the establishment of the diplophase refer to zygote variety.

Table 19

The regulation of recombination in eucaryotic systems

Regulatory factor	Recombination	
	Increasing	Decreasing
1. Chromosome number	High	Low
2. Chromosome type	Symmetrical (metacentric)	Asymmetrical (acrocentric– telocentric)
3. Chiasma characteristics	Random distribution High frequency	Localized distribution Low frequency
4. Structural hybridity	Absent	Present
5. Breeding system	Predominantly crossing	Predominantly selfing
6. Population size	Large	Small
7. Isolating mechanisms	Related species incompletely isolated	Related species strongly isolated
8. Length of generation	Short	Long

Note. After Grant (1958).

Gametic novelty depends in part on crossing-over, which leads to a recombination of linked genes, and in part on the random orientation of centromeres at the first and second divisions of meiosis, which leads to a recombination of unlinked genes. Zygotic novelty, on the other hand, is an expression of the genetic relationship of the mating parents, or more immediately, that of the gametes which fuse. The machinery of sexual heredity is thus designed to produce offspring with parental genes but not necessarily parental genotypes.

The effects of recombination, like those of mutation, are random relative to the needs of the individual and its offspring, so that variation is less likely to be useful as a population becomes more closely adapted to its environment. On the other hand, the greater the range of variation produced, the greater the chance of producing the new phenotypes required in new or altered environments. Once a new system of adaptation has been produced, however, the mechanisms of mutation and recombination can become an embarrassment in relatively stable environments. Even so, sooner or later the environment will change and then once again variation may be useful.

Short-term adaptation and long-term adaptability are thus in conflict, and success in terms of continuing survival depends on meeting their rival demands. Such a reconciliation can be achieved in two ways—by synthesis or by compromise. Let us see what they involve.

(a) *Synthesis.* A large variety of organisms, especially invertebrates and

many plants, have more than one reproductive mechanism. Standard
text-books of biology abound with cases where species effect rapid
colonization and maintenance by asexual or subsexual means in a
relatively constant environment, and resort to sexual reproduction when
conditions are unfavourable for simple vegetative growth. This alternation
may occur regularly and seasonally as in some aphids, or over much
shorter periods of time in simpler organisms whose environments can be
subject to rapid and extensive fluctuation. It may occur over longer
periods of time, too, as in many perennial plants where initial invasion
is by seed but subsequent colonization is by vegetative means. Other
species use a combination of in- and outbreeding (see p. 107) to effect a
similar reconciliation, achieving stability by inbreeding and variation by
outbreeding.

(b) *Compromise.* In sexually reproducing species, variation is a consequence of two things—meiosis and fertilization. It may, therefore, be
adjusted at two levels—by varying the level of recombination at meiosis
or by regulating the relationship between the gametes which are allowed
or enabled to fuse. Indeed some restriction on recombination operates
in all organisms simply because genes do not occur as free and independent entities but are always associated as chromosomes (linkage groups)
while individuals are segregated into more or less distinct breeding
units (mating groups) between which genetic exchange is difficult or
impossible. New combinations of genes are readily produced by outbreeding systems because the gametes which fuse are rarely immediately
related. But favourable genotypes are difficult to conserve under this
regime. Under conditions of inbreeding, on the other hand, fusion is
confined to gametes which are closely related in descent. This tends to
perpetuate the parental genotypes, and novelty is introduced only by
mutation and occasional lapses towards outbreeding.

   However, a measure of compensation can be achieved by combining
one kind of breeding system with the opposite type of meiotic system.
Thus, a meiotic system with restricted recombination can confer greater
stability on the outbreeder while a more open meiotic system improves
the variation potential of inbreeders.

   A convenient but not wholly adequate indication of meiotic recombination potential is given by the recombination index, which is the
sum of the haploid chromosome number and the average chiasma
frequency per nucleus. It represents the average number of segments
which are re-assorted at each meiosis. Clearly, the magnitude of this
index is affected in the same way and to the same extent both by changes
in the number of chromosome pairs and by changes in the number of
chiasmata. But the former is a numerical property of the karyotype
alterable only by chromosome mutation while the latter is a genotypically

controlled aspect of chromosome behaviour. Further while the first can
show only unitary change, the second can vary fractionally.

If there are n chromosomes in a haploid set, then random centromere
orientation at the first division of meiosis can yield a maximum of 2^n
gametic combinations of maternal and paternal homologues of which
$2^n - 2$ will be new. Thus, reducing the haploid number by one will lead
to a halving of the number of gametic combinations produced by the
orientation component of the meiotic system. When crossing-over is
also taken into account, the number of gametic combinations increases
to 2^r, where r is the recombination index, even if variation in the
distribution of the chiasmata are ignored. Thus a reduction of one in the
recombination index, by a change in either chromosome or chiasma
number, halves the number of gametic types and vice versa. In extreme
cases, chiasmata have been dispensed with entirely, an abolition which
requires the adoption of an alternative system for maintaining homolgue
association until co-orientation is assured. Obviously, chromosome
number alone determines the level of meiotic recombination in
achiasmate systems unless there is preferential segregation.

Differences in chiasma distribution affect the nature of the recombin-
ants rather than their range, and two main patterns are known in this
regard, distributed and localized. Chiasmata are said to be distributed
when they occur more or less randomly along the length of the chromosome irrespective of their over-all frequency which will vary somewhat
according to the length of the chromosome. Where there are differences
in length between chromosomes within the complement, the longer
chromosomes tend to form more chiasmata in total than the shorter ones,
but they generally have fewer chiasmata per unit length. This results from
the fact that all chromosomes, no matter how short, form at least one
chiasma. The disproportionately lower chiasma frequency per unit length
in the longer chromosomes itself constitutes a means of reducing recombination.

A more or less random distribution of chiasmata means that all parts
of the complement are subjected to the same rate of recombination. But
in some species this basic pattern is modified so that chiasmata are
virtually confined to certain regions irrespective of the length of the
chromosome. The net result of this is that substantial segments of the
chromosomes are hardly ever subjected to recombination. Segments of
this kind are described as being genetically differential. Thus, where
chiasmata are localized, different parts of the complement are recombined at very different frequencies. In some cases chiasmata are confined to pro-centric segments but more often they are restricted to
regions near the chromosome ends. The recombination index does
not accommodate this variable of the meiotic system nor does it take
into account structural hybridity. This is an alternative method of

creating differential segments to which we now turn.

Used in its broadest sense, the term mutation covers any permanent change in the hereditary constitution of an organism. As such it applies both to changes in genes and to changes in chromosomes. Chromosome mutations arise from errors of two kinds.

(a) Errors of chromosome replication or disturbances of chromosome movement during cell division can lead to changes in chromosome number. Numerical mutations of this kind may involve either single chromosomes (aneuploidy) or whole chromosome sets (polyploidy).

(b) Exchanges of non-homologous segments within or between chromosomes lead to changes in chromosome structure, and secondary numerical changes sometimes accompany or follow particular structural mutations. Three principal patterns of exchange are recognized (Fig. 35).

(*i*) *Inversions.* These are intrachromosomal rearrangements which lead to a reversal of the gene sequence in the chromosome segment concerned. Inversions fall into two categories: those in which the centromere is included in the inverted segment (pericentric) and those in which it lies outside the inverted region (paracentric).

(*ii*) *Interchanges.* These are interchromosomal exchanges in which the transposed segments may be large or small, equal or unequal. Two types are commonly found in nature. In one the terminal segments of two non-homologous elements are reciprocally exchanged (reciprocal translocation). The other involves essentially whole-arm exchange and is described as an unequal fusion.

(*iii*) *Centric exchange.* This involves either the dissociation of a two-armed chromosome into two telocentric elements as a result of breakage within the centromere (fission), or else an exchange which couples two telocentrics into one metacentric (fusion).

Exchange must always be preceded by chromosome breakage, because only broken ends are capable of re-joining. But the ends produced by breakage may not all reunite. In this event, segments which lack centromeres are produced and, being mechanically incompetent, such acentric fragments are invariably lost. Incomplete exchange therefore leads to quantitative deficiency and this often proves to be lethal, especially when large segments are involved.

The union of two centromere-containing segments, asymmetrical exchange, produces a dicentric chromosome which cannot be expected to be mechanically stable unless the two centromeres are so close together that they effectively function as one. For this reason stable karyotypes do not generally include dicentric chromosomes. As a rule, therefore, exchanges produce potentially viable chromosomes only when they are complete and symmetrical, and for the present purposes we need be con-

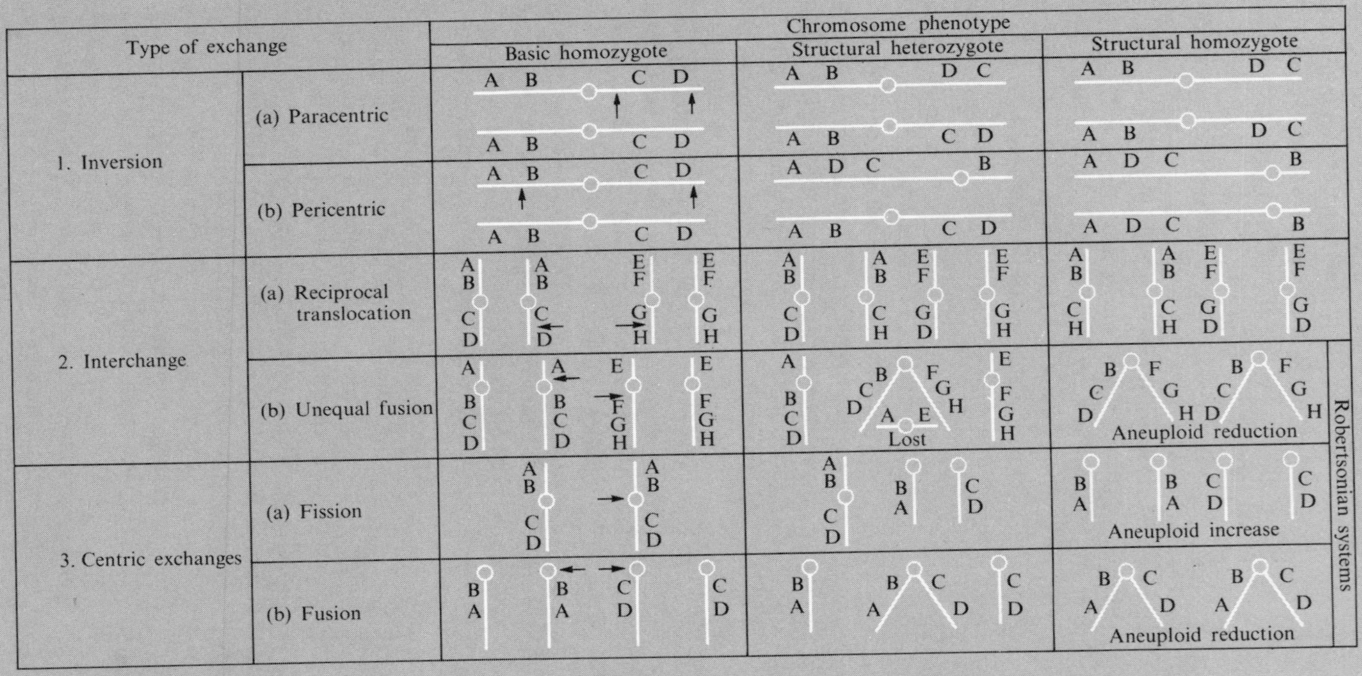

Chromosome phenotype

Type of exchange		Basic homozygote	Structural heterozygote	Structural homozygote
1. Inversion	(a) Paracentric	A B C D / A B C D	A B D C / A B C D	A B D C / A B D C
	(b) Pericentric	A B C D / A B C D	A D C B / A B C D	A D C B / A D C B
2. Interchange	(a) Reciprocal translocation	A B / C D / E F / G H	A B / C D / A E / C G H D	A E / C G H D / A E B F / C G H D
	(b) Unequal fusion	A B C D / E F G H	A B C D / A B C D / E F / G / D A E / Lost	B F / C G C G / D H D / Aneuploid reduction
3. Centric exchanges	(a) Fission	A B / C D	A B / C D / B A	C D / A B / B C / A D / Aneuploid increase
	(b) Fusion	B A / C D	B C / A D	B C / B C / A D / A D / Aneuploid reduction

Robertsonian systems

FIG. 35. Categories of structural chromosome mutation.

cerned only with rearrangements of this kind.

All structural mutations alter the linkage relationships of genes. In addition the amount and/or the position of effective crossing-over between linked genes may be reduced, restricted, or canalized. Like gene mutation, spontaneous chromosome mutation is a rare event. Thus when an exchange occurs it affects only one member of a pair of homologues so that the change is inevitably introduced into the complement in a heterozygous condition. That is, the mutant cell contains one normal and one rearranged member of a particular pair of homologues. Structural heterozygotes may, of course, subsequently inbreed or interbreed to produce individuals which are homozygous for the chromosome mutation (structural homozygotes). One needs, therefore, to distin-guish three distinct states in cases of exchange—the basic homozygote which has the two homologues in their original form; the structural heterozygote containing one unmodified and one modified member; and the mutant struc-tural homozygote, in which both homologues have been modified.

Mitosis in a structural heterozygote produced by complete sym-metrical exchange is perfectly regular, but meiosis is always modified. And the modifications of meiosis are important in relation to the role of structural changes in the regulation of recombination. Two funda-mental processes are operative during meiosis (see p. 47):

(a) the pairing of homologous chromosomes or chromosome segments at zygotene; and

(b) the subsequent formation of chiasmata by crossing over between specifically paired regions.

Both these processes occur not only in basic homozygotes and structural homozygotes but in structural heterozygotes too. However, structural hybridity affects the pattern of the first process and the con-sequences of the second. Consider, for example, the meiotic complications which are expected to arise in a paracentric inversion heterozygote. Com-plete homologous pairing in such a heterozygote is possible only if one or other of the relatively inverted segments forms a reverse loop (Fig. 36). Inversion-loop pairing can occur provided that the inversion is long enough, and this can be followed by crossing-over between the relatively inverted segments. A single crossover and certain combinations of double crossing-over in the inverted loop give rise to complementary duplication-deficiency chromatids which, in the case of paracentric inversions, are also either dicentric or acentric. The former often breaks during either first or second anaphase and the gametes which receive its fractured products necessarily contain an irregular gene constitution. The latter, since it lacks a centromere, fails to move and, unprotected by a nuclear membrane, it is soon destroyed.

The haploid chromosome complement contains one complete set of genes which work as an integrated whole in development. The development of haploid pollen in higher plants, for example, generally fails to proceed normally even if only small parts of the genome are not represented. Moreover, in both plants and animals, the capacity of a zygote to develop is seriously impaired by deficiencies from a balanced genome containing two or more entire sets of genes. Loss is more serious than gain though both alter the dosage of genes and lead to some disturbance in the pattern of development. Loss from diploids is more serious than loss from polyploids, which are buffered by the presence of more than two sets of chromosomes.

Most of the crossover products which arise following the recombination of genes within the limits of a paracentric inversion will give rise to unbalanced gametes and, if these survive, to unbalanced zygotes. Consequently, crossing-over within the limits of the inversion is expected to reduce the fertility of the heterozygote unless there is some system which compensates for its occurrence. In fact the high incidence of paracentric inversion heterozygosity in natural populations of various species of the genus *Drosophila* is associated with two compensating mechanisms—one in each sex. There is no crossing-over in the male meiotic sequence and so inversion heterozygosity does not introduce any complications at this stage of the sexual cycle. In the female, where crossing-over does occur, there is a preferential orientation and hence a preferential movement of non-crossover chromatids into the one functional meiotic product, for in female meiosis three of the meiotic products regularly abort as polar nuclei.

This preferential movement depends on the fact that when chromatid bridges are formed, they persist into the second meiotic division and tie together the centromeres of the dicentric crossover product. In so doing they ensure that a non-crossover chromatid passes preferentially to the innermost nucleus of the oocyte from which the functional female nucleus develops (Fig. 36).

One sees here how, at the level of individual nuclei, selective deaths can be debited to the account of non-selective elimination (see p. 117). In this way the preferential perpetuation of favourable gene combinations can be promoted with little or no reduction of fecundity.

There is some evidence that persistent bridges, formed during the chiasmate male meiosis of chironomids, hold the meiotic products together so that jointly they form a single, large, and non-functional sperm. By this means, eggs with balanced genotypes are protected against the sterilizing effects of the unbalanced products of male meiosis, for the fertility of females is a much more important component of fecundity. Short inversions and even comparatively long pericentric inversions may show linear pairing without reverse loops. Crossing-over in the inverted region is prohibited in this event because homologous regions are not aligned. For example, in a spontaneous

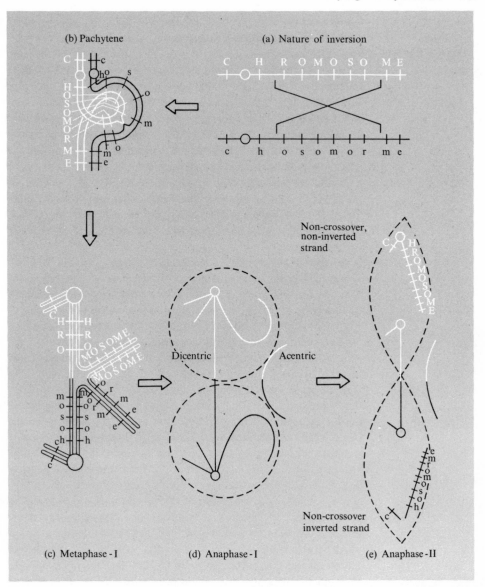

FIG. 36. Behaviour of a heterozygous paracentric inversion in female *Drosophila*.

paracentric inversion heterozygote of the grasshopper, *Camnula pellucida*, 261 out of 297 pachytene cells examined by Nur (1968) showed straight pairing of the relatively inverted homologues. Martin (1967) too has found extensive non-homologous pairing of inverted regions in male chironomids.

4.2. Chromosome polymorphism

4.2.1. Disruptive selection

Compensation mechanisms can explain why inversion heterozygotes in *Drosophila* are at little or no disadvantage compared to basic or inversion homozygotes, as far as fertility is concerned. But in themselves they cannot explain why inversion heterozygotes are present at such high frequencies in many natural populations of various species of *Drosophila*, which are polymorphic for a mixture of various inversion homozygotes and their heterozygous combinations. Individuals of *Drosophila* carrying inversions cannot be distinguished outwardly from those which do not. Indeed because of this it was originally assumed that the inversions were adaptively neutral and reached their incident frequency simply under the cumulative pressure of mutation. There is now good evidence, however, that inversions have adaptive properties which means, amongst other things, that while the investigator may not be able to distinguish the different types of flies phenotypically, the flies themselves can. This evidence is of two kinds.

(a) Different inversion sequences show regional variation with regard to both occurrence and relative frequency. The latter also shows seasonal variation within populations. Thus the polymorphism appears to allow a given population to respond to the changing character of the local environment both in space and in time.

(b) Under laboratory conditions, alternative inversion sequences can be shown to respond differentially in relation to the levels of such external agencies as temperature and relative humidity.

Upon what then do such advantages depend? For the most part, the functional female meiotic products of inversion heterozygotes are those in which recombination has not occurred within the limits of the inversion. Genes within the inverted block thus form a closely linked system which is sometimes referred to as a supergene because all the loci within the inversion are inherited as a unit. In other words the inverted gene block behaves as a differential segment. For this reason inversions can serve to hold together coadapted groups of alleles, that is groups of alleles which produce high fitness because of specific interactions between them. And different inversions can, of course, hold distinct co adapted complexes which arise in relation to specific selection patterns. Different inversion complexes can thus become adapted to different facets of the local environment. What is more, an individual which is heterozygous for two such complexes may be better buffered against environmental flunctuations, because it contains two integrated and internally balanced gene complexes each of which is able to cope with a different set of environmental conditions. In other words, the polymorphism itself is adaptive because it allows the *population* to respond *genetically* to spatial and temporal trends and fluctuations in the environment in a more efficient way than is possible with chromosomal monomorphism. Further, heterozygosity is itself adaptive because it enables hybrid *individuals* to adjust *phenotypically* to environmental variation in a more effective way than is possible with chromosome homozygosity. Indeed, not only must a high level of heterozygosity always attend polymorphism in outbreeding systems but heterozygote advantage will always generate polymorphism.

The pattern of selection involved in the production of chromosome polymorphism is called disruptive selection, because it splits a population into distinct classes each of which has a different adaptive potential. It is, we may say, a selection for diversity, the various karyomorphs performing different adaptive roles at different times in the same population, or at the same time in different populations. The simplest and best-known example of such a situation is, of course, the sex-chromosome mechanism which serves to isolate male and female sex-determining genes within one species. Here there are two morphs which are mutually dependent because they are mutually adapted to serve quite distinct roles in the population.

4.2.2. *The conservation of heterozygosity* Having considered how recombination at meiosis can be modified by structural changes, let us give some consideration to the manner in which restrictions may also operate on the second process which regulates recombination in the sexual cycle, namely fertilization.

The extent to which fertilization is effective in achieving recombination between whole nuclei obviously depends on the average genetic relationship of the gametes which fuse. This, in turn, usually reflects the genetic relationship of the mating parents. The factors which influence these relationships collectively constitute the breeding system. For convenience it is customary to distinguish two extreme systems of breeding—inbreeding and outbreeding. Inbreeding refers to the situation where progeny are produced from one parent or from a pair which are closely related. Outbreeding describes matings between individuals which are not immediately related.

The breeding system actually defines the effective size of the gene pool, that is the genetic reservoir of the breeding group from which individuals receive their genotypes in one generation and to which they contribute genes in the next.

The effective size of this pool is determined by properties of two kinds, one relating to the nature of the individuals, the other to the nature of the population. Thus it is affected by both the way in which mating is organized within the group (mating system) and the effective size of the group within which mating occurs.

For example, there may actually be more gametic recombination, and hence a higher level of heterozygosity, in a large population with a high incidence of inbreeding than in a small population where cross-fertilization is obligatory. Indeed, obligate crossing may be of little consequence even in a large population recently derived from a small number of progenitors. Thus, while the mating system which operates in a breeding group can be ascertained with little difficulty, the breeding system may be difficult to define.

In theory, inbreeding is expected to lead to homozygosity at numerous genetic loci, and where there is no heterozygosity there can be no effective recombination. But while inbreeding tends to fix a particular allelic combination in the homozygous state, its characteristic products may become homozygous for different combinations of alleles. This tends to produce a series of distinct inbred lines which, if crossed, can rapidly regenerate heterozygous combinations.

Inbreeding is thus characterized by two inter-related effects. The first is a strong tendency to induce the differentiation of a previously large, randomly breeding population into a series of smaller subpopulations, sometimes referred to as demes. Secondly, as the various subpopulations are differentiated, the amount of genetic variability within each small subpopulation becomes reduced as homozygotes increase at the expense of heterozygotes. However, the theoretical curves which describe this descent to homozygosity are unrealistic because they are based on the assumption that the genotypes of one generation make a contribution to the next generation which is strictly in proportion to their numbers, that is, the effects of selection are ignored. But this assumption is not valid because unaccustomed inbreeding is invariably accompanied by a fall in vigour and, while this does not strictly follow the descent to homozygosity, it affects homozygotes more than heterozygotes.

Consequently, the heterozygotes of any one generation tend to make a greater contribution to the next generation than their more homozygous contemporaries. For this reason the level of heterozygosity does not actually fall as quickly as theory predicts. This differential fitness probably underlies the remarkable genetic variability which is often found even within natural populations of inbreeding species. This suggests that high variability is essential to the survival of many populations. It also suggests that a change in any component of the variation-producing system which leads to an undue restriction of variability is frequently accompanied by compensatory changes in other components of the system, which lead to a conservation of heterozygosity.

Maintaining the level of heterozygosity by the selective elimination of homozygotes is very uneconomical, especially when the change to inbreeding is abrupt. Fitness is better served by systems which decrease the frequency of production of those individuals which selection is inclined to eliminate. We see this principle elegantly demonstrated in the case of interchange heterozygotes. Interchange hybridity, like inversion hybridity, is a system of linkage. Interchanges, however, achieve a level of linkage which cannot be obtained by either intrachromosomal rearrangements or a mere reduction of chiasma frequency, since otherwise non-homologous chromosomes become linked in their transmission although they are physically discontinuous. Thus interchange hybrids are an exception to the exception to Mendel's second law—genes on different, non-homologous chromosomes do not always assort independently and so the haploid chromosome number can exceed the number of linkage groups.

An interchange disturbs the neat pairwise homology of the chromosomes and creates a situation in which each of the chromosomes involved shares partial homology with two others. This provides a basis for multiple formation and the valency of the potential multiple is equal to $2(i + 1)$ where i equals the number of overlapping heterozygous interchanges. Thus, a single interchange leads to a multiple association of IV, two overlapping interchanges produce a multiple of VI and so on (Fig. 37).

Theoretically, the centres of the crosses formed by pairing in interchange heterozygotes should mark, and be defined by, the points at which the exchange event occurred, i.e. the points where linear homology is interrupted. But, as in other cases of structural hybridity, non-homologous or interrupted pairing is not uncommon in the vicinity of a change in linear homology. The morphology of the associations formed at meiosis in interchange hybrids depends on the patterns of pairing and of chiasma formation and the manner of multiple orientation. Fertility too is affected by these aspects of chromosome behaviour.

In this regard, three conditions must be satisfied. First, all the members of the potential multiple must become associated into a single configuration, for otherwise their regular disjunction relative to each other cannot be achieved consistently. Clearly, this requires pairing and chiasma formation in all or all but one of the arms of the pachytene cross. In the former event, the chromosomes of the multiple will become associated in a closed ring while in the latter event a chain will be formed.

Second, the necessary chiasmata must be confined to the pairing segments and excluded from those which lie between the centromeres and the interchange points. Crossing-over in these so-called interstitial segments results in the attachment of two incompletely homologous chromatids to a common centromere. This presents orientational problems which cannot be resolved and is incompatible with fertility in excess of 50 per cent (see below).

Thirdly, the multiple must orient in such a way that the two non-interchanged members of the multiple pass to one pole and the two interchanged ones pass to the other (alternate orientation). In rings this requires a twist at or before metaphase I to produce a figure of eight. In chains it involves a zig-zag arrangement of the centromeres. Not only do these arrangements produce balanced gametes but they are the only

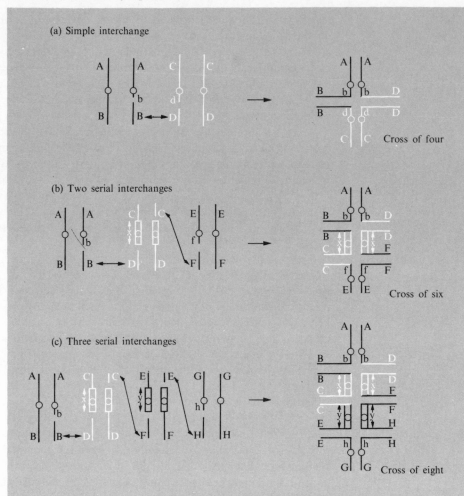

(a) Simple interchange

(b) Two serial interchanges

(c) Three serial interchanges

FIG. 37. Zygotene configurations produced by interchange heterozygotes.
Note: b, d, f, and h represent interstitial segments while x and y constitute
differential segments.

patterns of orientation which lead to the production of exclusively
balanced gametes (Fig. 38).

If a chiasma occurs in one of the interstitial segments, then no pattern
of orientation can lead to regular fertility, even when rings or chains of
four are formed. The reason for this is that such a chiasma automatically
leads to the movement of interchanged and non-interchanged chromatids
to the same pole at anaphase I (chromatid non-disjunction) and this leads
to unbalance in half of the products of meiosis, the crossover or the non-
crossover chromatids passing to the viable gamete depending on the

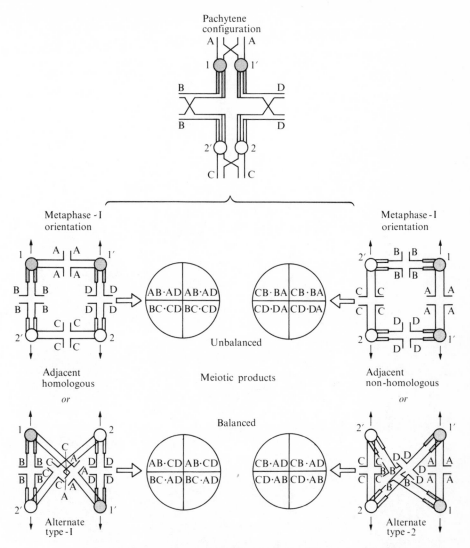

FIG. 38. The consequences of meiosis in a ring of four interchange heterozygote when crossing-over is confined to the pairing segments. The unlikely diamond-shaped (discordant) orientation is not considered.

orientation (Fig. 39). A similar situation obtains if crossing-over occurs in both interstitial segments, but disjunction then depends on second division orientation. That unbalanced products do lead to zygotic inviability has been shown most convincingly by the study of Jaylet (1971) on the transmission of an induced interchange in the amphibian *Pleurodes* (Fig. 40).

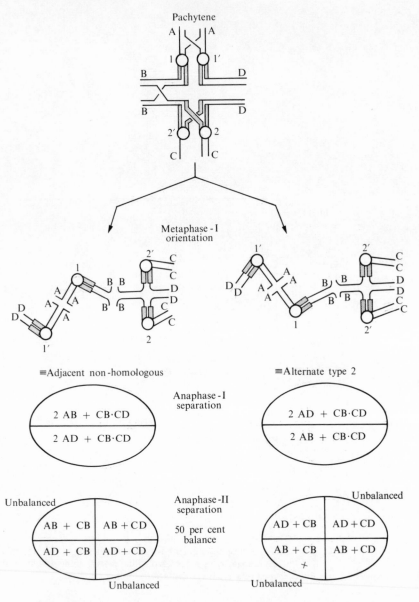

FIG. 39. The consequences of crossing-over in the interstitial segment of an interchange heterozygote. Note: the two types of orientation are mechanically indistinguishable.

In multiple associations of six or more, a third type of segment can be distinguished, namely that between the two points of interchange (Fig. 37). Such a region consists either of the sum of adjacent interstitial

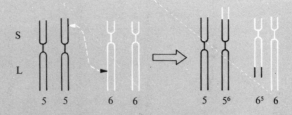

Constitution of zygote	Dosage	Developmental effect
2 (5) ; 2 (6) 5, 5⁶ ; 6, 6⁵ 2 (5⁶);2(6⁵)	Normal—balanced	Normal
2 (5); 6,6⁵ 5,5⁶; 2 (6⁵)	Triplicate for end of 5S, deficient for end of 6L	Lethal
5,5⁶; 2(6) 2 (5⁶) ; 6⁵,6	Triplicate for end of 6L deficient for end of 5S.	Subvital
2 (5) ; 2 (6⁵)	Quadruplicate for end of 5S deficient for end of 6L	Gastrula aborts
2 (5⁶); 2 (6)	Quadruplicate for end of 6 L deficient for end of 5S	

FIG. 40. Consequences of zygotic imbalance in an X-ray-induced interchange in *Pleurodes* (after Jaylet (1971)).

segments (in which case it includes a centromere) or else of the difference between overlapping interstitial segments (in which case it does not contain a centromere) depending, respectively, on whether the two interchanges occurred in opposite arms or the same arm of the common chromosome. Crossing-over in differential segments reduces fertility to the same extent and for the same reason as crossing-over in interstitial segments. But it has another, unique consequence for it produces chromosomes which differ in respect of their end combinations from those of the original heterozygote. If viable gametes containing chromatids which are crossovers for differential segments fuse with viable gametes which are not crossovers for the differential segment, the resulting offspring will have a smaller ring than the parent. In effect, therefore, differential segment crossing-over reverses an interchange and leads to a breakdown of large rings.

To enjoy high fertility, therefore, interchange hybrids must avoid recombination in the interstitial and differential segments. That is, these segments must be genetically differential. In fact in all the known successful interchange systems not only do the interstitial and differential segments function as supergene blocks but, in addition, the chiasmata are localized distally in the pairing segments. Consequently, the genetically defined differential segments can be extended to include those regions of the cytologically defined pairing segments which are proximal to the point of chiasma formation (Fig. 41).

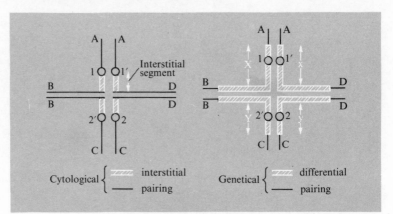

FIG. 41. Effect of chiasma localization on the extent of the genetically maintained differential segments in an interchange heterozygote.

In terms of viable, balanced gametes, therefore, the consequences of inbreeding an interchange heterozygote can be represented as in Table 20. The progeny are of two kinds, non-ring and ring, in equal numbers. With respect to the genetically differential segments, the non-ring offspring will be genetically homozygous while the ring types will be as heterozygous as the parents. By contrast, crossing two structurally similar heterozygotes with dissimilar ancestry gives ring and non-ring progeny, both of which are highly heterozygous. Interchange heterozygosity can thus serve to regulate the distribution of genic heterozygosity. More particularly, it leads to a bimodal distribution under conditions of inbreeding, promoting both extremes at the expense of intermediate levels of hybridity. Consequently, selection for allelic heterozygosity under such breeding conditions frequently favours structural heterozygosity too. Levin, Howland, and Steiner (1972) have recently carried out an analysis of allozyme polymorphism in a large population of the evening primrose, *Oenothera biennis*. This species, of course, is the classic case of complete and permanent interchange heterozygosity, for all its chromosomes are serially related by a system of interchanges. In consequence one giant ring multiple is formed at meiosis incorporating every chromosome in the complement. Levin and his colleagues conclude from their study that, while the proportion of polymorphic loci is no greater than in a normal meiotic system, *Oe. biennis* does show an unusually high proportion of heterozygous loci per individual, and five of the nineteen loci they examined showed fixed and permanent allelic heterozygosity.

Table 20

*The consequences of inbreeding versus crossing in interchange hetero-
zygotes (see also Fig. 39)*

Inbreeding		
		Eggs
Viable gametes 1–2 and 1′–2′	1–2 AXB + Cyd	1′–2′ AxD + BYC
Sperm 1–2 AXB + Cyd	AXB Cyd ‖ + ‖ AXB CyD Non-ring	BYC − CyD ∣ ∣ AXB − AxD Ring
1′–2′ AxD + BYC	BYC − CyD ∣ ∣ AXB − AxD Ring	AxD BYC ‖ + ‖ AxD BYC Non-ring

Products: Non-ring: homozygous for differential segments
Ring: heterozygous for differential segments

Crossing		
		Eggs
Viable gametes 1–2 and 1′–2′	1–2 AB + CD	1′–2′ BC + DA
Sperm 1–2 bc + da	bc − CD ∣ ∣ AB − da Ring	BC DA ‖ + ‖ bc da Non-ring
1′–2′ ab + cd	AB CD ‖ + ‖ ab cd Non-ring	BC − cd ∣ ∣ ab − DA Ring

Products: Non-ring: heterozygous for differential segments
Ring: heterozygous for differential segments

Interchange heterozygotes form a regular part of the genetic system
in at least seven of the 19 genera in the family Onagraceae, namely
Clarkia, Stenosiphon, Camissonia, Calylophus, Gaura, Gayophytum,
and *Oenothera.* Moreover, the last four genera include species in which
all the chromosomes are united at meiosis in one giant ring. In some of
these, entire populations consist of permanent, true-breeding hybrids,
since balanced lethal combinations of alleles have developed which cause

the death of zygotes which result from the union of similar gametes. Since recombination is excluded from genetically differential segments, recessive lethal mutations can accumulate in them secondarily. If the four chromosomes of a translocation heterozygote are represented as 1·2, 3·4, 1·4, and 3·2 such that 1·2 and 1·4 carry a $+_1\ell_1$ arrangement in their interstitial segments while 3·2 and 3·4 carry a $+_2\ell_2$ arrangement, where ℓ_1 and ℓ_2 are recessive lethal mutations, then only two classes of gamete can form, namely $+_1\ell_2$ or ℓ_1+_2. Any zygotic combinations leading to homozygosity for one or both of the lethals will die and only the heterozygote $\ell_1+_2/+_1\ell_2$ can survive. Such balanced zygotic lethals then enforce permanent chromosomal heterozygosity (Fig. 42).

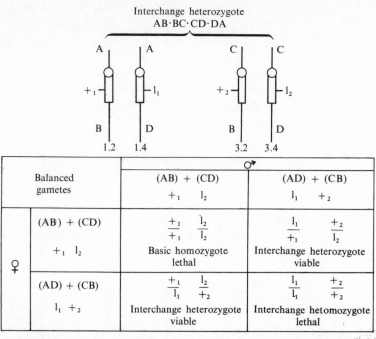

	$\male$	
Balanced gametes	(AB) + (CD) $+_1$ l_2	(AD) + (CB) l_1 $+_2$
$\female$ (AB) + (CD) $+_1\ l_2$	$\dfrac{+_1 \quad l_2}{+_1 \quad l_2}$ Basic homozygote lethal	$\dfrac{l_1 \quad +_2}{+_1 \quad l_2}$ Interchange heterozygote viable
(AD) + (CB) $l_1\ +_2$	$\dfrac{+_1 \quad l_2}{l_1 \quad +_2}$ Interchange heterozygote viable	$\dfrac{l_1 \quad +_2}{l_1 \quad +_2}$ Interchange hetomozygote lethal

FIG. 42. Maintenance of permanent interchange heterozygosity by balanced lethals.

The maintenance of heterozygosity by this means, however, involves a 50 per cent zygotic lethality even when linkage is complete. But this can be avoided by complementary gametic elimination. Thus, if a plant is hybrid for two complexes, one containing a pollen lethal (ℓ_p), the other a megaspore lethal (ℓ_e), both in the differential regions of the ring, it can be represented as $\ell_p+/+\ell_e$. Such a hybrid would form two kinds of meiotic product ℓ_p+ and $+\ell_e$. The first class would not survive in pollen because it contains the pollen lethal; only $+\ell_e$ pollen would be produced. In the ovules, on the other hand, only ℓ_p+ gametes would survive.

Apart from producing early rather than later death, gametic lethals in themselves do little to improve the biological economy of hetero-zygote preservation by homozygote elimination. But the megaspore lethals are associated with a system of megaspore competition known as the Renner effect. As a result of this the functional megaspore in each ovule is that containing the pollen lethal and so the selective system can operate while full ovule fertility is maintained. The Renner effect depends on the fact that in higher plants, as in animals, only one of the four meiotic products is functional on the female side (non-selective elimin-ation). Where megaspore lethals are present, two of these are subject to selective death but their place is taken by one of the meiotic products from which the megaspore lethal has been excluded by segregation. Here too, therefore, selective elimination is at the expense of the non-selective (see pp. 96 and 104).

These principles can be extended to multiple translocation systems too since irrespective of size of the ring involved, its members fall into two alternating groups of chromosomes. Each group acts as a unit at meiosis in the sense that all chromosomes of the group always pass to the same pole. Thus in *Oenothera muricata* six translocations require seven chromosomes to move together in one group and seven in another, and the species is kept permanently heterozygous for the two groups by means of two complementary gametic lethal complexes. The a-complex acts as a pollen lethal, while the α-complex contains a megaspore lethal (Table 12).

Interchanges are not the only means of conserving heterozygosity. Consider the possibilities afforded by parthenogenesis. Genetic systems in which reproduction is entirely parthenogenetic and in which all popu-lations are female have often been divided into two types—automictic (meiotic) and apomictic (ameiotic). In the former, some form of meiosis occurs but a complementary doubling of chromosome number takes place at some stage to prevent the formation of reduced gametes. In the latter meiosis has been suppressed altogether and there is no need for compensatory doubling because there is no reduction. Rather, meiosis is replaced by an essentially mitotic division where one of the two products becomes the egg nucleus while the other forms a polar body or its plant equivalent. Here all sibships would of necessity be isogenic and would retain the maternal genotype. In cases where this is hetero-zygous then clearly the heterozygosity would be conserved.

In neither type of parthenogenetic system is there any recombination of genes between different individuals, but in some automictic types there is at least the possibility of genetic segregation provided the parent is heterozygous (cf. inbreeding). Even here, however, the system

in fact embraces a number of distinct genetic mechanisms. In fact
compensatory doubling may be achieved in three different ways.

(a) *Pre-meiotic doubling followed by a normal meiosis.* This can result
either from a failure of anaphase separation at the last pre-meiotic
mitosis, or else by endoreduplication during pre-meiotic interphase.
Unless the various genomes in the original type are represented in the
hemizygous state (i.e. ABC) then pre-meiotic doubling results in the
production of an essentially autopolyploid state (e.g. AA $\rightarrow$ AAAA) in
the mother cells. This polyploidization is over and above any polyploidy
which may obtain initially. Thus, while such a doubling (a kind of
internal fertilization) can provide a means of avoiding reduction, it
potentially introduces all the meiotic problems associated with even
levels of polyploidy (see p. 150). These can be avoided in various ways,
for example, by the restriction of metaphase association to pairs of
chromosomes. This can occur presumably because homologues are
differentially distributed in pairs ($\equiv$ sister chromatids) in the pre-meiotic
and, hence, zygotene nuclei. Consequently they show preferential pairing
in twos. This pairing is preferential also in that it occurs between what
were either the sister chromatids of the pre-meiotic mitosis, or else
sisters of pre-meiotic replication. Thus, although crossing-over occurs,
it has no genetical consequences; the chiasmata simply serve to hold
homologues together until their orderly separation can be accomplished.

For example, as White (1970) has shown, *Moraba virgo* ($2n = 15$) is
a diploid orthopteran consisting only of females. The mechanism of
parthenogenesis involves an endoreduplication in the oocyte nucleus
prior to meiosis, followed by synapsis between sister chromosomes, that
is pairing is restricted to endoreduplication products which form fifteen
pseudobivalents. Meiosis, therefore, restores the egg nucleus to a genetic
condition which is exactly equivalent to that of the pre-endoreduplicated
oogonial cell, provided of course that no mutation has occurred. Thus
any heterozygosity in the mother can be maintained, or even increased
by mutation. Indeed, in all the 19 localities from which *M. virgo* has
been collected, it has proven to be heterozygous for a number of chromosomal rearrangements. What is more, at two of the localities, additional
rearrangements have been superimposed on the 'standard' heterozygous
karyotype and these must have arisen subsequent to the establishment
of parthenogenesis. Here then we have a mechanism which not merely
conserves heterozygosity but actually augments it.

In fact, it is worth bearing in mind that although sexual reproduction
is often discussed in relation to the maintenance of heterozygosity,
mutations are heterozygous at the time of their origin. What is more,
they would remain in this state were it not for sexual reproduction. In
other words, it is the formation of homozygotes which is completely
dependent on sexual reproduction because the probability of achieving

this state by coincident or sequential mutation is so prohibitively low that the prospect can be ignored.

To take another example, North American salamanders belonging to the *Ambystoma jeffersonianum* complex have been shown by Uzell (1963) to be parthenogenetic triploids ($2n = 3x = 42$). Two parthenogenetic biotypes are known. One, *A. tremblayi,* combines a haploid set of *A. jeffersonianum* with two sets of *A. laterale,* while the other, *A. platineum,* has two sets of *A. jeffersonianum* with one of *A. laterale.* In both of these triploids 42 bivalents are present in oocyte nuclei. Consequently, there must be a pre-meiotic doubling, $42 \rightarrow 84$, followed by bivalent formation. Triploid eggs, produced by a $6x \rightarrow 3x$ reduction, need to be stimulated by sperm from either of the two sexual diploids before parthenogenetic development can proceed. This is called gynogenetic development.

On the other hand in the parthenogenetic stick insect *Carausius morusus,* Koch, Pijnacker, and Kreke (1972) have shown that there is an extra replication of DNA at a despiralized stage immediately following pachytene of female meiosis.

(b) *Normal meiosis followed by fusion of the first or second cleavage nuclei.* Since cleavage nuclei are haploid and, barring mutation, genetically identical, restoration of the unreduced number by their fusion must lead to immediate homozygosity at all loci. This mechanism which is known to occur in scale insects and whiteflies (*Hemiptera: Homoptera*) thus rigidly enforces complete homozygosity.

(c) *The suppression of one of the meiotic divisions.* There are two possibilities in this connection:

(*i*) central fusion mechanisms involving the fusion of first division products; and

(*ii*) terminal fusion either by the suppression of the second division or else by secondary entry of the second polar body and its re-fusion with the egg nucleus.

The genetic consequences which follow central and terminal fusion are shown in Fig. 43 and Table 21. Where chiasmata are distributed (see p. 100) both systems will eventually give the same result. But where chiasmata are localized there will be a clear difference. Considering only one chiasma per bivalent, the suppression of first division has similar consequences to the random fusion of meiotic products, so far as the distal segments are concerned. But for loci in the proximal segment, transmission is essentially mitotic. With suppression of second division, on the other hand, heterozygosity is conserved for distal segments. The similarity between this system and that already seen for the differential segment of interchange systems is at once obvious and clear.

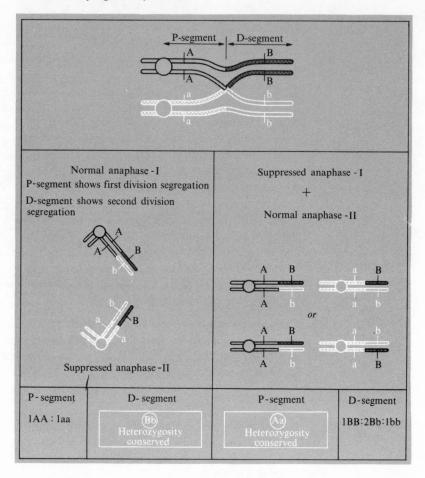

FIG. 43. Genetic effects of the suppression of first or second division of meiosis on proximally and distally sited loci. Note: conservation of heterozygosity is facilitated by a terminal localization of crossing-over with first division suppression and by pro-centric localization when the second division is suppressed.

Of course, if chiasmata are distributed freely along the chromosome, then eventual homozygosity for all loci is inevitable whether reduction is avoided by suppressing the first or second of the meiotic divisions. What is more, unless the regular occurrence of a single proximally localized chiasma can be guaranteed, an unlikely prospect, homozygosity for all loci will be quickly reached if reduction is avoided by suppressing second division. With first-division suppression, however, the number of distally located chiasmata is irrelevant to the conservation of heterozygosity in proximal segments. This then would appear to be the safer system of meiotic suppression to operate.

Table 21

The genetic consequences of sub-sexual parthenogenesis

Mechanism	Consequences
1. Suppression of AI	Heterozygosity conserved for loci showing pre-reduction therefore no variation in respect of them.
	Heterozygosity reduced by one-half in each generation for loci showing post-reduction. Variation released in respect of them for as long as heterozygosity lasts.
2. Suppression of AII	Heterozygosity conserved for loci showing post-reduction therefore no variation in respect of them.
	Heterozygosity lost immediately and completely for loci showing pre-reduction. Thus heterozygosity for them is expressed as variation once and for all.
3. Fusion of post-meiotic mitotic products	Complete and immediate homozygosity for all loci. Thus any potential variation in the heterozygous state is immediately expressed as free variation.

Thus, from a genetic point of view, the significant distinction in parthenogentic forms is not between auto- and apomictic systems but between systems that enforce homozygosity as opposed to those that facilitate the conservation of heterozygosity. Where parthenogenesis is facultative, it offers a conservative mode of reproduction to complement the more innovative sexual system. Obligate parthenogenesis, on the other hand, is expected to obtain where sexual reproduction presents difficulties. Such difficulties are of two main kinds.

First, actually effecting sexual reproduction may present problems owing to impediments at either the meiotic or the fertilization stage of the cycle. The former are likely to exist in hybrids and polyploids, especially uneven polyploids. The latter may well occur in cross-fertilizing species which are dioecious, self-incompatible, or which rely on an external agent, such as an insect, for pollination. Significantly, apomixis is most common among hybrids, polyploids, and erstwhile outbreeders. For example, 13 polyploid races have been found in some 26 species of earthworm and 10 of these polyploids are obligatorily parthenogenetic. Weevils too, show extensive apomixis coupled with polyploidy. Thus of 37 parthenogenetic races belonging to 15 genera, 23 are triploid, 9 are tetraploid, 4 are pentaploid, and only one is diploid (Table 22).

Second, even when sexual reproduction can be achieved its products may be selected against. Thus enforced inbreeding may produce un-balanced and unwanted homozygotes while wide outbreeding may give excessive variation. Apomixis, most forms of which conserve hybridity

Table 22
Weevil species which have reproductively or cytologically different races

	Bisexual diploid	Parthenogenetic			
		Diploid	Triploid	Tetraploid	Pentaploid
Otiorrhynchus niger	+		+	+	
O. dubius	+			+	
O. scaber	+		+	+	
O. subdentatus			+	+	
O. salicis	+		+		
O. rugifrons	+		+		
O. gemmatus	+		+		
O. chrysocomus	+		+	+?	
Peritelus hirticornis			+	+	
Trachyphloeus bifoveolatus	+		+		
Barynotus squamosus	+		+		
B. moerens			+		+
Polydrosus mollis	+	+	+		
Liophloeus tessulatus	+		+		
Catapionus gracilicornis				+	+

Note. After Suomalainen (1969).

and, correspondingly, limit variation, can be useful under these circumstances also. Of course the adoption of parthenogenesis is not without problems. Thus parthenogenetic forms not uncommonly produce a high percentage of inviable progeny, presumably due to imperfections in the genetic system. Thus errors of fusion, or the occurrence of crossing-over in unusual positions, could well lead to the formation of genotypes homozygous for recessive lethals or sublethals. Nevertheless subsexual reproduction, like other modifications in the pattern of the sexual cycle, can serve to conserve those favourable allelic combinations which the sexual process has created but which by the very same mechanisms of meiosis and fertilization it tends also to destroy.

Finally, in this closed kind of reproductive system, let us notice that heterozygosity and variation, far from going together, present themselves as alternatives. If the potential variability of heterozygotes is released as free variation, the heterozygosity is lost and its restoration must wait on mutation.

4.3. Flexible systems Success in colonizing new habitats depends initially not on systems of conservation but rather on the presence of flexibility, for new habitats are usually opened up only locally and sporadically and are often available for colonization for only a relatively short time. In angiosperm plants, Ehrendorfer (1965) has distinguished three principal types of colonizers (Table 23). All three represent optimum modes of balance between flexibility and stability. It is to these that we now turn.

Table 23
The three principal types of colonizers in angiosperms

Character	Type 1	Type 2	Type 3
1. Life form	Perennial	Annual	Annual
2. Reproduction			
(a) Sexual	Crossing	Selfing	Crossing but self-incompatibility rarely complete
(b) Vegetative	Often present	None	None
3. Chromosome characters			
(a) Number	Often polyploid	Diploid or polyploid	Diploid often with aneuploid reduction
(b) Symmetry	Relatively symmetrical	Variable	Relatively asymmetrical
4. Population structure			
(a) Isolation	Predominantly geographical and ecological	Predominantly floral and genetic	Predominantly genetic
(b) Differentiation	Mostly allopatric	Partly sympatric	Partly sympatric
(c) Hybridization	Often extensive in contact areas	Reduced or absent despite contact	May be absent despite contact
(d) Variability	Populations relatively variable	Populations relatively uniform	Populations relatively uniform
5. Examples	*Achillea millefolium* complex $(2x, 4x, 6x)$ *Chrysanthemum leucanthemum* $(2x, 4x, 6x, 8x)$ *Gallium anisophyllum* $(2x, 4x, 6x, 8x, 10x)$ *Gallium pusillum* $(2x, 4x, 6x, 8x)$	*Martricaria martricoides*, *Anthermis arvensis*, *Pterocephalus plumosus*	*Artemisia*, *Scabiosa brachiata* *Cephalaria syrica*, *Cruciata articulata*

Note. After Ehrendorfer (1965).

There can be little doubt that the restoration or improvement of
fertility represents the major evolutionary role of polyploidy in
hybrids. But the real significance of the allopolyploid lies not in the
state of polyploidy as much as in the inter-genomic hybridity generated
by bringing together the genotypes of two distinct species. Allopoly-
ploidy thus allows for a measure of recombination between interspecific
genotypes. The precise measure, however, depends on the behaviour of
the allopolyploid in question. In a number of allopolyploids only
bivalents are formed, so that they behave meiotically as diploids and
show essentially disomic inheritance. Indeed such behaviour may occur
even when the chromosomes brought together in the hybrid are known
to he homoeologous, i.e. to show partial homology. Homoeologous
chromosomes may show varying degrees of homology according to the
number of shared homologous loci and their locations. Further, as in
the case of truly homologous chromosomes, these loci may be occupied
by identical alleles or by different alleles of the same gene. Pairing
between chromosomes carrying identical alleles is called homogenetic
or homogenic, while that between chromosomes carrying dissimilar
allelic forms is called heterogenic. Thus even in amphidiploids, with
exclusively bivalent pairing, homo- and heterogenic pairing can and
must be distinguished.

There are also segmental allopolyploids in which multiples are formed
which include both homologues and homoeologues. Thus in a number of
allotetraploid grasses quadrivalents form with a high frequency and show
a significantly greater incidence of alternate orientation at metaphase I
than would be expected on a random basis. Some workers have interpreted this to be a means of securing a regular 2:2 separation of the
constituent chromosomes. Clearly this is not an adequate explanation
since adjacent orientation of a ring quadrivalent guarantees this. What
significance then can one attach to such behaviour?

Consider an allotetraploid multivalent of the type $aB.aB.B^1c.B^1c$,
where B and B^1 represent homoeologous segments. If we compare the
efficiency of random bivalent formation in such a system with that of
random quadrivalent orientation, there is clearly no basis for establish-
ing any preference. Both give 75 per cent gametes with the balanced
$aB.B^1c$ combination necessary to maintain the tetraploid (Table 24).
On the other hand, consistent alternate orientation not only ensures
balanced gametes but it also ensures recombination between the
homoeologous segments brought into the hybrid from the two parents.
Only preferred bivalents with heterogenic pairing will do this and only
one half of these give balanced products. Directed orientation of ring
quadrivalents is thus important for the production of recombination
products.

Allopolyploids are frequently expected to arise as single founder

Table 24
Recombination in segmental allotetraploids

(B and B$'$ = homeologous segments)

(a) If the two patterns of bivalent pairing are at random, 75 per cent of the gametes are balanced (aB.B$'$c) but intergenomic recombination occurs in the production of only one third of these so that only 25 per cent of all gametes are balanced and recombinant. Further, while fertility is favoured by homologus pairing, intergenomic recombination requires homoeologous pairing.

(b) With multiple association, 75 per cent of all gametes are again balanced (aB.B$'$c) but consistent alternate orientation would ensure that all gametes were both balanced and recombinant.

$$
\begin{array}{ccc}
\text{aB} & & \text{B}'\text{c} \\
\| & \times & \| \\
\text{aB} & & \text{B}'\text{c}
\end{array}
\quad \text{Crossing}
$$

$$\downarrow$$

$$\text{aB}\quad \text{B}'\text{c} \qquad \text{F}_1 \text{ diploid hybrid}$$

$$\downarrow \qquad \text{Chromosome doubling}$$

$$
\begin{array}{cc}
\text{aB} & \text{B}'\text{c} \\
\text{aB} & \text{B}'\text{c}
\end{array}
\qquad \text{Allotetraploid}
$$

(a) *Bivalent formation*

(i) Homologous pairing

$$
\left(
\begin{array}{cc}
\text{aB} & \text{B}'\text{c} \\
\| & \| \\
\text{aB} & \text{B}'\text{c}
\end{array}
\right)
$$

(ii) Homoeologous pairing

$$
\begin{array}{cccc}
\text{aB} & \text{aB} & \text{aB} & \text{B}'\text{c} \\
\| & \| \text{ or } \| & & \| \\
\text{B}'\text{c} & \text{B}'\text{c} & \text{B}'\text{c} & \text{aB}
\end{array}
$$

Random orientation

(b) *Multivalent formation*

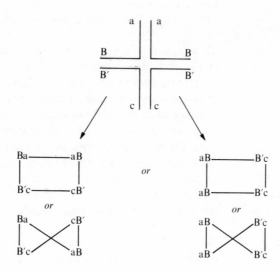

individuals. Such individuals may find an opportunity for sexual reproduction through incomplete self-incompatibility or various forms of asexual or subsexual propagation may be available to them. Here flexibility is primarily due to the recombination of the initial hetero-zygosity through heterogenic pairing of chromosomes from different parental genomes coupled with the possibility of additional hybridiz-ation with races of the same or different ploidy levels. Stability may arise by selfing or through the adoption of a non-sexual method of propagation. Stability with sexual reproduction can also be achieved by homogenetic or preferential pairing.

These polyploid perennial colonizers thus draw predominantly on the genetic diversity built up at the diploid level but they recombine and mobilize this diversity. For example, as Zohary (1965) has shown, the recent evolutionary success of the weedy polyploid *Aegilops* is attribut-able not to polyploidy as such but to the fact that polyploidy made possible the establishment of a large, common gene pool. Indeed the genetic material of the various diploid groups, which are completely isolated, could only be brought together and remoulded at the poly-ploid level. Such genetic systems are admirably suited for rapid colonization. Whereas diploid species in this genus have a limited range of morphological variation, the tetraploids and hexaploids are extraordinarily variable. Again, while the diploids show sharp specific boundaries, those between the polyploids are blurred, and mixed colonization is common throughout the Middle Eastern countries. In addition, the species build-up at the polyploid level has been far from random, but has rather followed the most successful trends previously established at the diploid level. The most successful of the diploids have contributed the pivotal genomes for the polyploid clusters. The other diploids have provided the material for recombination.

4.3.2. *Outbreeding annuals*

These constitute a second category of colonizers. In such cases a high rate of outcrossing produces variable offspring which have good chances for survival under conditions of expansive colonization. These forms often show asymmetrically constructed chromosomes, a consequence of structural changes leading to an aneuploid reduction series. This produces a marked structural differentiation of the karyotype and tends to subdivide and fractionate the effective breeding group which makes the swamping of new types by hybridization less likely. Recombination frequency also tends to fall with increasing asymmetry, partly by the reduction in chromosome number and partly by a reduction in the chiasma frequency of the chromosomes.

4.3.3. *Inbreeding annuals*

Here again single founder individuals may initiate the sequence of colonization. The short generation time makes for variability. So too does the fact that the breeding system of many predominantly self-pollinated species often provides for a rapid alteration in their levels of outcrossing. Even a low level of outcrossing between co-existing homozygous genotypes in a population allows a large number of new genotypes to be produced in a limited number of generations. In some cases this outcrossing may be combined with the production of allopolyploids between different selfing lines. Stability, of course, follows from the selfing system. Thus in several senses the evolutionary strategy of these mainly autogamous annuals combines many of the virtues of the first two types.

DeWet (1971) has proposed a novel means of producing flexibility through reversible tetraploidy. Haploids derived from diploids, and polyhaploids derived from amphiploids are sterile. Where tetraploids combine the genomes of closely allied, but differently adapted, intraspecific populations, however, polyhaploids can expose masked gene combinations to selection. In *Dicanthium* these neo-diploids are often better colonizers than their agamic parents. They also serve for an introgressive exchange at the tetraploid level by providing an indirect means for the agamospecies to maintain contact with sexuality.

4.4. Chromosome change and speciation

So far we have discussed those features of population structure which result in the stabilization of each population in its own particular environment while, at the same time, maintaining a level of variability so that change can be accomplished as conditions require. Such a pattern of evolutionary change is termed anagenesis. Here, as its environment alters with time, a species undergoes changes that maintain or improve its adaptedness but, nevertheless, it continues to act as a single species.

A second major category of evolutionary change is termed cladogenesis. This involves the formation of a new species and depends on the development of isolating mechanisms which restrict gene flow and gene exchange. Thus, while interbreeding holds individuals together in populations, the lack of it leads to fractionation and the separation of species.

Isolation is most simply achieved by spatial separation, and the most common basis for speciation is generally held to be through eco-geographical isolation. In becoming adapted to different environmental conditions following such separation, populations may become genetically differentiated as races or subspecies. By extension this process may lead to the races becoming so different genetically that they can no longer interbreeed. Thus:

$$\text{populations} \longrightarrow \text{races} \longrightarrow \text{species,}$$

and according to this allopatric hypothesis, populations become

genetically different in the process of adapting to different environments. Certainly, geographical and ecological isolation are in themselves
often sufficient to interupt gene flow and so initiate race formation.
When genetic divergence has gone so far that gene exchange between
such races becomes disadvantageous, then it is assumed that selection
will promote the development of reproductive isolating mechanisms
should spatial contact be re-established.

It is a matter of empirical observation that even the most closely
related species are often found to differ in chromosome number or
chromosome structure or in both. This has led to the suggestion that
in such cases these differences have played a primary role in the
speciation process itself. This is not to say that speciation cannot occur
in the absence of chromosome change. For example, the chromosomes
in certain species of *Drosophila* are known to be homosequential, that
is, they do not differ at all in the banding sequences of their polytene
chromosomes. These are unquestionably good species because they
co-exist in the same habitat without interbreeding. This ability to co-
exist sympatrically without losing identity through hybridization is
regarded by evolutionary biologists as of prime importance in recogniz-
ing distinct species.

Not only can speciation occur in the absence of chromosome change
but, as we have already seen, many chromosome mutations have nothing
to do with speciation. Rather they are involved in producing a polymorphism within a given species and, in fact, help to maintain the
integrity of the group. Even where related species do differ in karyotype,
this does not in any sense prove that the chromosome change caused
the speciation. Such change in itself could be simply an incidental
accompaniment of the divergence which can be expected to follow
successful isolation. Nevertheless, despite all such arguments, karyotype
changes can play a crucial role in speciation since they offer one means,
and a very direct means, of producing hybrid sterility. Under appropriate
conditions such a post-mating barrier to gene flow can become rein-
forced by suitable pre-zygotic mechanisms. Thus chromosome barriers
possess an additional feature which is not shared by eco-geographical
barriers. They may directly stimulate the development of secondary
mechanisms of isolation. As such they can, in theory, function as fairly
strong primary genetic isolating mechanisms. Further, chromosome
mechanisms leading to reproductive isolation can be efficient at their
conception and they need not improve on contact. They can, therefore,
develop quite effectively in isolated populations, for their efficiency
depends on errors at meiosis. These errors arise from disturbances in
orientation and segregation which are sometimes, though not always,
associated with a loss of pairing efficiency.

For example, when species are hybridized, either in nature or in the

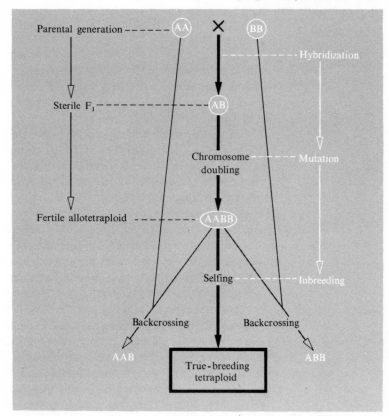

FIG. 44. Restoration of hybrid fertility by allotetraploidy and inbreeding.

laboratory, the F_1 produced is often infertile because of disturbances in pairing. Such an infertile F_1 can, however, be converted into a self fertile but intersterile F_2 by tetraploidy (Fig. 44). And the production of such a tetraploid does not depend on exophenotypic selection, for the mutation almost automatically improves fertility. Notice also that this mechanism works under sympatric situations.

The same argument can be applied in the case of structural chromosome mutations, with one added complication. Here the mutations which are important as species barriers are those which lead to a reduction in the efficiency of meiosis in the heterozygous state. Thus those which behave regularly at their inception are not likely to serve as effective isolating mechanisms because their efficiency in this connection depends on the meiotic inefficiency of the heterozygous state.

Consider, for example, the case of a chromosome interchange which, like other mutations, originates in the heterozygous state. Where the heterozygote enjoys high fertility and is otherwise favoured by selection, it can lead to the development of a stable polymorphism. Where, on the

other hand, the initial mutant heterozygote is ill-behaved at its inception
it has the potential for forming an efficient isolating mechanism if—and
only if—it can be quickly converted into, and then rapidly fixed as, a
structural homozygote. To achieve this state the initial mutant would
have to pass through a bottleneck since structural homozygotes would
occur far too rarely in large populations to have any chance of becoming
established.

Clearly, individuals heterozygous for a structural chromosome mutation
that greatly reduces fertility can have little chance of contributing to the
next generation in a normal population, let alone affecting the chromosome
composition of the entire population. The potential for forming an
efficient isolating mechanism can only be realized in cases where the
initial heterozygote is rapidly converted into the structural homozygote
despite its inefficiency. This is most likely in small populations, whether
these represent founder or relict populations. Here novel conditions of
selection can be expected and a measure of inbreeding is inevitable. If
an individual of low fertility is removed from competition with normal
members of the parental population then its degree of infertility is of
less importance, provided it can produce some progeny.

Whereas speciation involving a gradual accumulation of gene differences
is expected to be a slow and progressive process, speciation involving
chromosome change could be rapid and need not represent an extension
of eco-geographic race formation. What is more, the barriers resulting
from geographical separation are usually incomplete and remain so for
indefinite periods of time. Those resulting from chromosome change,
on the other hand, are expected to be much more decisive. Indeed it is
not without significance that where chromosome differences do occur
between species, populations intermediate in chromosome arrangement
are rarely found.

A further and perhaps more important significance of structural
chromosome rearrangements may lie not so much in their possible role
in effectively isolating individuals as in the part they can play in virtually
isolating homologous segments even when they share a common nucleus.
Thus, as we have seen, structural hybridity imposes considerable restraints
on gene exchange so that of the variety of assortment patterns which
are possible between a chromosome and its altered homologue, only a
few lead to effective recombination. Now, if even homologous segments
are restrained from exchanging their mutational experiences, they are
free to diverge within the constraints imposed by selection. Those genic
differences they show initially, however extensive, will tend to be
maintained and, eventually, they will be added to.

Inter-homologue isolation is initiated when the rearrangement arises,
that is, in the initial structurally heterozygous mutant, and it operates
in the heterozygotes of all subsequent generations, including those

which may arise later by the crossing of the alternative homozygous forms.

The magnitude of this isolation potential can be judged from the extreme differences which can be generated between X- and Y- chromosomes even though the divergence of sex chromosomes is constrained by the compelling need for mutual adaption. This argument can, of course, be extended to the phenotypic differences between the sexes themselves. And these are often as great as those commonly regarded as large enough to justify specific designation.

While genetic variability remains in large outbred populations, its potential for changing a species may not be realized. When, however it is dispersed into small populations, unexpected and novel properties may emerge. For example, a number of strains of the normally bisexual species *Drosophila mercatorium* show an extremely low frequency of female parthenogenesis. In most cases less than one egg in a thousand develops without ferilization. But by selection and the sequestering of this variation in small laboratory populations Carson (1967) has been able to produce stocks with up to a 60-fold increase in the frequency of parthenogenesis. Inbreeding coupled with novel selection is thus a powerful force for dispersion and fixation. Again, at a single locality, a high altitude Swiss Valley—the Val Poschiavo—a dark-coloured mouse has been known since 1869 when it was first described under the specific name of *Mus poschiavinus*. Recently it has been shown that whereas *Mus musculus*, the house mouse, has 40 telocentric chromosomes, *Mus poschiavinus* has $2n = 26$ with seven pairs of metacentrics, 5 pairs of telocentric autosomes, and a pair of telocentric sex (XY/XX) chromosomes. F_1 hybrids between *M. poschiavinus* and *M. musculus* have a somatic set of $20 + 13 = 33$ with 7 metacentrics and 26 telocentrics, and maximum meiotic association leads to the formation of 7III + 5II + XYII. Less than one half of the second metaphase figures examined in such hybrids are euploid. Not surprisingly, therefore, this F_1 interspecific hybrid shows a marked reduction in fertility. The 7 fusions for which *M. poschiavinus* has become homozygous have thus built up an effective barrier which isolates this species from its progenitor *M. musculus* even when contact and mating between them is enforced.

It is important to emphasize that the genetic changes which build up isolating mechanisms can arise quite independently of those which produce clearly recognizable phenotypic differences between species. When this happens, sibling species result; these are species which are so much alike that they are almost impossible to distinguish from each other on conventional taxonomic criteria. For example, *Drosophila miranda* is a distinct species from *D. pseudoobscura* in genetical terms. Morphologically, however, they are so similar that taxonomists find

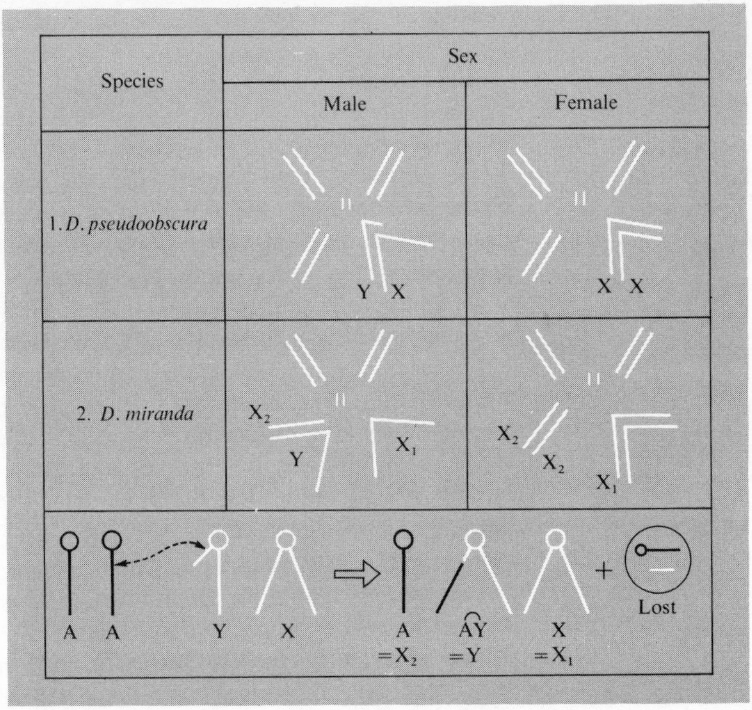

FIG. 45. Chromosome comparisons between two sibling species of *Drosophila*.

them difficult to distinguish. At the chromosome level they are readily distinguishable because there is a marked difference in their sex-chromosome system (Fig. 45). Experimental hybrids between the two forms are sterile and the male products in particular are thoroughly abnormal. A detailed comparison of the chromosomes in the hybrid is possible in this case because, although meiosis is abnormal and cannot be studied, homologous chromosomes pair also in the polytene salivary gland nuclei. The pairing here is an exaggerated and enhanced form of the somatic pairing found in mitotic tissues of dipterans. A comparison of the pairing and banding patterns of the two parental forms, in the salivary gland chromosomes of the F_1 hybrids, shows that the chromosomes have in fact been completely restructured as a result of numerous inversions and interchanges. As a consequence of this the chromosomes no longer pair effectively at meiosis and gamete formation, in turn, breaks down.

The kinds of chromosome change which come to constitute intraspecific polymorphisms in the members of a group are often very different from those associated with their divergence. This is is not surprising. The nature of a balanced polymorphism is often such as to render unlikely, or even preclude, the establishment of a monomorphic condition. This is most clearly evident in the case of sex dimorphism but it obviously applies to other systems too. For example, over 200 paracentric inversions are known within the many species of *Drosophila* that have been studied cytologically. By contrast the number of instances of intraspecific pericentric inversion and centric fusion can be counted on the fingers of one hand. Yet these two latter categories of change are among the most common of the interspecific differences in the genus. The general point is that a polymorphic situation will obtain where a condition of mutual dependence exists. And the longer the morphs stay together the more mutually adapted, and dependent, they become. Consequently after a long association they are no more likely to be successful separately than are the members of other symbiotic systems.

4.5. Chromosome races

There is a third evolutionary strategy which chromosomes can adopt, which is in some respects intermediate between their role in maintaining balanced polymorphisms within populations and that in isolating species from one another. This is in the development of chromosome races homozygous for one or more structural rearrangements, the most common involving whole-arm transfers, or Robertsonian transformations. Two mechanisms have been proposed which could lead to such Robertsonian relationships, namely the fusion of acrocentric or telocentric entities and the fission of metacentrics (Fig. 35). And both have been claimed to account for Robertsonian situations in nature (Table 25).

Fusions automatically lower the number of linkage groups, so reducing the possibility for interchromosomal recombination. Fission has the opposite effect because it increases the number of linkage groups. It may also lead to an increase in intrachromosomal recombination, since each telocentric must then form at least one chiasma to secure segregation, whereas in a metacentric a single chiasma can serve to maintain the bivalent. Both fusion and fission agree in that the rearrangement is effective in modifying linkage relationships, though not equally so, in both the heterozygous and the homozygous states.

All such cases of Robertsonian races have three common properties:
(a) there is a sequential relationship between the present geographical distribution of the races (Fig. 46);
(b) the races show contiguous allopatry; there are few, if any, localities in which the different races are sympatric and few or no natural hybrids are known except in parapatric situations where narrow hybrid zones are formed;
(c) migration and gene flow between races is minimal, presumably because ecological differences form barriers to dispersal.

Such situations have been considered by some workers as intermediate evolutionary stages between a single polymorphic population

Table 25

Some examples of chromosome races in mammals and arthropods

Species	Racial karyotypes 2n	Total No.	Presumed type	Natural hybrids	Reference
Mammals					
Spalax ehrenbergi	52, 54, 58, 60	4	Fission (+ pericentric inversion)	Rare 2n = 53 and 59	Wahrman, Goitein, and Nevo 1969
Thomomys bottae	76	40	Reciprocal translocation or pericentric inversion or added blocks of heterochromatin?	None described	Patton 1973
Thomomys talpoides	40, 44, 46, 48, 56, 58, 60	8	Not known, possibly fusion + pericentric inversion. Forms with 2n = 48 show interpopulation variation	None described	Thaeler 1968
Perognathus goldmani	50, 52 54 60	6	Fusion (+ pericentric inversion)	Rare, present at 2 localities but no sympatric contract zones	Patton 1973
Spermophillus richardsoni	34, 36	2	Fusion/fission unspecified	2n = 35 hybrid at narrow zone of potential contact	Nadler, Hoffman, and Greer 1971
Rattus rattus	38, 42	2	Fusion (+ pericentric inversion)	2n = 40 and 39 in South Pacific	Yosida, Tsuchiya and Moriwake 1971
Sorex araneus	Race A 20 + XY₁Y₂ Race B 18 + XY₁Y₂ (but polymorphic for Robertsonian changes in	2	Races differ minimally by 3 pericentric inversions and one tandem translocation	Races overlap without hybridization at two localities in Switzerland	Ford and Hamerton 1970

Species	Racial karyotypes 2n	Total No.	Presumed type	Natural hybrids	Reference
Arthropods					
Jaera syei	18, 20, 22, 24, 26, 28 (♂)	6	Fusion/fission unspecified	18 × 20, 18 × 26 and 24 × 26 hybrids synthesized experimentally	Lecher 1967
Didymuria violescens	26, 28, 30, 31, 35, 37, 39 (♂)	10	Fusion	Rare hybrids at overlap zones and others synthesized experimentally	Craddock 1970
Dichroplus pratensis	15, 16, 17, 18, 19, (♂)	9	Fusion	None described	Saez and Mosquera 1971
Podisma pedistris	23♂ (X0) 24♀(XX), 22♂(neo-XY) 22♀(neo-XX)	2	Fusion	Not known	Hewitt and John 1972

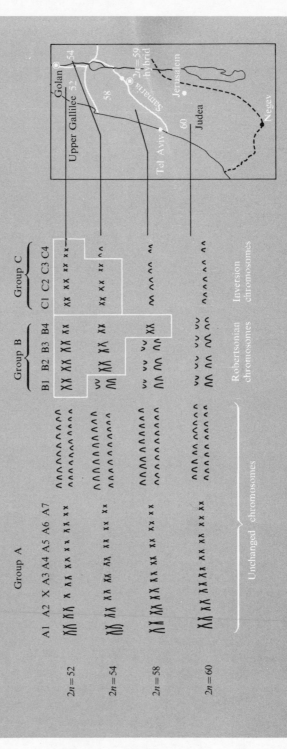

FIG. 46. The four chromosome races of *Spalax ehrenberghi* and their distribution in Israel.

and that of reproductively isolated species with distinct monomorphic karyotypes. Such an explanation assumes that, on division of the polymorphic population, heterozygotes are selected against and eliminated so as to leave chromosomally distinct populations. However, an initially large population polymorphic for so many differences is as difficult to conceive as is a process which would split such a population into so many races. Added to this is the fact that the Robertsonian heterozygotes in question are often incapable of serving as effective chromosomal isolating barriers because the trivalent, formed between the one metacentric and the two rods, is pre-adapted in showing regular segregation. In consequence little loss of either viability of fertility is expected on meiotic grounds.

To take one example: as Craddock (1970) has shown, the phasmatid *Didymuria violescens* includes a spectacular array of chromosome races which are geographically distinct. Hybrids have been found in narrow parapatric contact zones and these show high fertility, as too do laboratory-reared synthetic hybrids between the different races. Indeed this may well have been one of the factors which led to the relatively rapid fixation of the chromosome condition in the homozygous state. Such trivalents thus appear to have an inbuilt drive mechanism.

This is not to suggest that Robertsonian systems cannot or have not been used for the establishment of intraspecific polymorphisms or for species isolation. The former condition is found in the shrew *Sorex araneus* where the diploid number varies from 20 to 32 and no less than 6 autosomal elements can be represented by alternative metacentric or twin telocentric forms, though no more than three have been found to vary within a single population. In this case Ford and Hamerton (1970) are of the opinion that heterozygote advantage forms the basis of a balanced polymorphism. The latter condition has been described by Smith (1966) in the twice-stabbed ladybird beetle *Chilocorus*. Here *C. tricyclus* ($2n = 20 = 3$ ring II + 6 rod II + neo-XY) is confined to the interior of British Columbia and the adjoining portion of Washington, while *C. hexacyclus* ($2n = 14 = 6$ ring II + neo-XY) is known only in Saskatchewan and Southern Alberta. In a population at the head of Crowsnest Valley on the immediate east side of the Rocky Mountains, the two species co-exist—though with *C. hexacyclus* predominating—together with F_1 hybrids and hybrid derivatives with intermediate chromosome numbers of which $2n = 17$ forms are the most common. These intermediate forms regularly contain one, two, or three Robertsonian trivalents which are frequently linearly oriented. Segregational sterility is most severe in the triple heterozygotes with three trivalents which are also the most common class, and Smith estimated the over-all sterility in the hybrids to be about 40 per cent. Because of this, F_2 progeny are rare. It appears that *tricyclus* adults

invade the resident *hexacyclus* community year after year, carried by
the prevailing westerly winds that funnel through the Crowsnest pass.
Despite this they are constrained from successfully invading *hexacyclus*
territory by the infertility and lack of fecundity of the hybrids which
are formed. Nevertheless it would appear that back-cross derivatives of
these hybrids to *hexacyclus* may well lead to an introgression of
tricyclus genes into the *hexacyclus* genotypes.

Clearly much depends on the influence which Robertsonian changes
have on meiotic behaviour and hence on fertility. In most European
domestic mice, *Mus musculus*, and all Asiatic and sub-Asiatic species
there are forty telocentric elements in the chromosome complement.
An extensive array of fusions have, however, recently been demonstrated
in populations taken from alpine valleys in Switzerland and from
neighbouring central European areas. This involves predominantly
fusion homozygotes but occasional heterozygotes do occur (Table 26).
All individual autosomes with the exception of pair number 18 appear

Table 26

Chromosome variation in alpine populations of Mus musculus

Population	$2n$	Constitution
1. Birsfelden	39	Heterozygous for one fusion
2. Haldenstein	39, 38	Heterozygotes and homozygotes for one fusion
3. Val Bregaglia	40, 39, 38	Polymorphic for one fusion
4. Bottmingen	38	Homozygous for one fusion
5. Chiavenna	35	Homozygous for two and heterozygous for a third fusion
6. Lorrach	33	Homozygous for three and heterozygous for a fourth fusion
7. Val Mesolcinna	28	Homozygous for six fusions

Note. Data from Gropp, Winking, Zech, and Miller (1972).

to take part in the formation of at least one, and sometimes more than
one, fusion metacentric. And, of particular importance, considerable
differences exist in the frequency of irregular meiotic disjunction
between the different fusion heterozygotes produced in experimental
hybrids between these races.

Chromosome races may also develop in conjunction with other
structural rearrangements too, but these are as yet less well understood.
A very striking example is that report by Patton and Dingman (1969) in
the pocket gopher *Thomomys bottae*. Here extensive interpopulation
variation in chromosome structure has been observed but all individuals
show $2n = 78$. The most readily distinguishable difference between

population karyotypes lies in the relative number of two-armed and one-armed chromosomes in the complement. It is not known whether the observed morphological differences are due to pericentric inversion, translocation, or added blocks of heterochromatin, but more than forty separate karyotypes have been recognized to date although the species has been sampled over only half of its known range (Patton 1973).

Of course, some of these races may well prove to be incipient species. Four chromosome forms of *Spalax ehrenbergi* exist with karyotypes of 52, 54, 58, or 60 chromosomes. These are distributed clinally and parapatrically from north to south Israel. These may well be sibling species, since mating trials reveal that aggression is more pronounced and success in copulation is lower when inter-racial crosses are attempted under laboratory conditions. Certainly the same kinds of differences which exist in chromosome races can also be found between acknowledged species of the same genus as is evident in the case of *Rattus*. This genus is distributed primarily in south-east Asia from where it has spread throughout the world, partly by its own capacity for movement and migration, and even more so in recent times by commercial traffic. Today this genus is represented by an impressive world-wide array of sub-species and species.

Yosida (1973) has shown that the black rat, *Rattus rattus*, which is the presumed progenitor of all other types, is itself subdivided into two races. Black rats collected from various localities in east and south-east Asia, ranging from Japan to Java, have $2n = 42$, with 13 telocentric and 7 metacentric pairs together with a telocentric $XY\male/XX\female$ pair. In addition autosomes 1, 9, and 13 show polymorphism for pericentric inversions. On the other hand black rats collected in Oceania (New Guinea, Australia, and New Zealand) have $2n = 38$, with two groups of large metacentrics derived from the fusion of autosomes $\overline{4\ 7}$ and $\overline{11\ 12}$ of the Asian black rat. Twelve additional species of rat collected in Asia and Oceania have been analysed cytologically and these fall into three groups (Fig. 47).

Group 1. Species with $2n = 42$ which are closely related to the Asian black rat, namely, *annandalei* (b) in Fig. 47, *exulans* (c), *muelleri* (d), and *norvegicus* (e).

Group 2. Species with numbers lower than $2n = 42$ which involve fusion, namely, *bowersii* ($2n = 40$; $\overline{11\ 12}$ (f)), *fuscipes* ($2n = 38$; $\overline{4\ 11}$, and $\overline{9\ 12}$ (h)), *leucopus* ($2n = 34$; $\overline{2\ 12}$, $\overline{3\ 6}$, $\overline{4\ 7}$, and $\overline{5\ 10}$ (i)), and *conatus* ($2n = 32$; $\overline{2\ 3}$, $\overline{4\ 7}$, $\overline{5\ 6}$, $\overline{8\ 9}$, and $\overline{10\ 12}$ (j)).

Group 3. Species with fewer small metacentrics, namely, *sabanus* ($2n = 42$; 2 pairs of metacentrics, 19 and 20 (k)), *canus* and *huang* ($2n = 46$; 3 pairs of metacentrics, 18, 19, and 20 (m)), and *niviventer* ($2n = 46$; 2 or 3 pairs of small metacentrics).

Clearly, speciation in Group 2 rats has involved an extension of the process of fusion which has also occurred in the racial divergence of *Rattus rattus*.

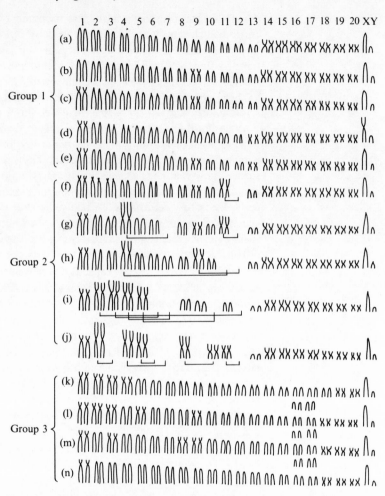

FIG. 47. Karyotype variation in 14 species of the genus *Rattus*. The bars indicate the source of the fusion metacentrics (after Yosida (1973)).

(a) *R. rattus tanezumi* (2n = 42)
(b) *R. annandalei* (2n = 42)
(c) *R. exulans* (2n = 42)
(d) *R. muelleri* (2n = 42)
(e) *R. norvegicus* (2n = 42)
(f) *R. bowersii* (2n = 40)
(g) *R. rattus rattus* (2n = 38)
(h) *R. fuscipes* (2n = 38)
(i) *R. leucopus* (2n = 34)
(j) *R. conatus* (2n = 32)
(k) *R. sabanus* (2n = 42)
(l) *R. canus* (2n = 46)
(m) *R. huang* (2n = 46)
(n) *R. niviventer* (2n = 46).

This example, and others like it, raises the problem of the relationship between gene change and chromosome change in the origin of species. The Norway rat, *R. norvegicus,* for example, corresponds closely to one of the racial karyotypes of the black rat and similarly retains the inversion polymorphism in autosome 13. In addition, however, it is also polymorphic for inversions in autosomes 3 and 11. Norway rats also show a pattern for transferrin (a serum protein) similar to one of the types (TfN) present in Japanese black rats. By contrast, in the Oceanian type black rat the transferrin pattern is markedly different and, of course, these animals are homozygous for two fusions. Nevertheless although on both grounds Norway rats appear to be more closely related to Asian black rats than to Oceanian black rats, no hybrids have ever been produced between the Norway rat and the Asian black rat either in nature or in the laboratory, whereas Oceanian and Japanese black rats do hybridize with one another both in the laboratory and in nature.

Finally, it is unlikely that all these examples of racial divergence will prove to represent speciation in transit. In at least some of these cases there are good grounds for believing that the present pattern of partially isolated populations will persist as a complex of races with restricted gene flow operating across inter-racial boundaries.

5 Conclusion—the chromosome as a genetic mechanism

A hierarchy is an organization with successive levels of authority, each level being in some degree subordinate to that above it. The effect of such stratified subordination is to constrain and integrate the activities of each system so that they function coherently. Biological systems are characterized by their order and this order comes in hierarchic steps. Hierarchy in biology thus means principally a hierarchy of order. For example, the genome of an organism is hierarchically organized with quanta of heredity which are correlated in their statistical behaviour at meiosis, since they are organized into linkage groups which behave in an integrated manner. Our present conception of this hierarchy is a linear sequence of nucleotide bases, 3 of which make a codon, 300 or so of which make a cistron, 2000 or more of which make a chromosome, two or more of which make a genome. In addition, each chromosome has a unique geometry which regulates both its mechanics and its dynamics. Thus the chromosome is more than a simple linear array of independent genetic units. The geometry of the genes it contains determines particular patterns of interaction. Not only is the chromosome part of the hereditary hierarchy, but it has a hierarchical organization all of its own.

This hierarchy of order is also a hierarchy of information, for biological order is produced on the basis of pre-coded information. The morphological changes in which chromosomes engage reflect differential patterns of chromosome organization and chromosome activity, and serve to mark biochemical events which relate to the release of information within cells, and its transfer between cells or generations. Whether this information is to be transcribed or repressed, replicated or reposited, recombined or conserved is thus a direct reflection of the state of the chromosome. Indeed most of these events depend on an interaction between different components of the chromosome hierarchy. Thus the low-level order code at the nucleotide level is supplemented by higher-level components, since the nucleotide sequence can specify only single-step relationships and these have none of the integrating capabilities required by a system of epigenetic control. This implies that the DNA code is not itself always sufficient to permit function, but that an accessory organization is also required. In this connection, the information implicit in significant positional relationships among genes appears to be an important component of hierarchical organization. In *Salmonella* and *Escherichia* for example, genes whose enzymic products are involved in the biosynthesis of a given amino acid are often closely linked. Thus the histidine cluster involves nine genes and no gene unconcerned with histidine synthesis is included in it. The genes of the cluster are under a common control mechanism and constitute an operon. The same state of affairs holds for the smaller constellations of genes involved in the synthesis of leucine and tryptophan. Such extensive clustering of genes, however, occurs only in the enteric group of bacteria.

A number of considerations suggest that the genes of an operon have a common origin. The fact that the enzymes of the operon form a reaction chain implies overlapping specificities which, in turn, suggests some measure of structural homology and common ancestry. In addition, the phenomenon of allosteric inhibition, in which the product of the last enzyme of the chain inhibits the activity of the first, is consistent with the idea that the pathway evolved by retro-evolution and that the first enzyme carries a stigma of its origin. Indeed it is interesting that the genes controlling the first and the last of the enzymes involved in histidine synthesis are found side by side on the *Salmonella* genome, and a comparable condition has been found in the leucine operon. If these clusters evolved backward by tandem duplication followed by functional differentiation through mutation, then the map order implies that the gene controlling the last step in both operons is the one that gave rise to the gene controlling the first step since the latter is located at the end of the map. The only conceivable alternative is that the individual genes of the operon originated in different regions of the chromosome and were transposed to their present position by structural rearrangement.

The pathways of leucine, tryptophan, and histidine synthesis are the same in *Neurospora* and bacteria, but in *Neurospora* the genes that control them are largely scattered among the chromosomes. This scattering of functionally related genes may well be connected with the development of the eucaryote chromosomal apparatus. Whether this means that the evolution of a multiple chromosome mechanism brought with it a more elaborate genetic control system which dispensed with the need for operons, or whether operons would be basically unstable in eucaryotes because of the enhanced possibilities for unequal crossing-over is not clear. Certainly unequal crossing-over would be far less important in predominantly asexual bacteria.

It is true that the dispersion of related genes need not necessarily prevent their unitary control. Thus the 8 genes of the arginine system in *E. coli* are not organized in one cluster but they are controlled in parallel, each locus presumably having its own copy of a regulatory element. The amount of information which a system can contain or transmit can, of course, be increased by increasing the redundancy of the system, and it is significant that many eucaryotes show a marked redundancy in the organization of the genome in the form of many families of closely related if not identically reiterated sequences of DNA. In addition, eucaryotic systems include a component, hetero-chromatin, which has no known counterpart in procaryotes and which plays a key role in gene regulation. Some of the reiterated DNA sequences are known to occur in heterochromatic areas but since heterochromatin rarely constitutes more than 20 per cent of the genome, it is clear that not all repeat DNA can occur in heterochromatin. In exceptional cases, the amount of heterochromatic material can be increased vastly. For example, eighty of the ninety-three species of the subgenus *Drosophila* known in Hawaii have a haploid mitotic karyotype consisting of five rods and a dot chromosome. The other species with $n = 6$ have added heterochromatin. In some this is present only on the dot chromosome but in the case of the giant species *D. cyrtoloma* all six chromosomes have acquired an extra heterochromatic arm and it is significant that in this species about 60 per cent of the total DNA is satellite.

As yet we have no clear idea what significance can be attached to the varied patterns of heterochromatin which are known in different eucaryote species. Of possible importance to this problem is the issue of unstable genetic loci which have been uncovered in flowering plants. This work rests on the initial discovery by Rhoades (1936) that a recessive a_1-allele in maize, which had not previously been observed to mutate, changed to a dominant A_1 form with very high frequency if another factor, termed dotted, was introduced into the same genome. Subsequently McClintock (1965) showed that unstable loci relate to a more general hierarchical control system with the following properties.

(a) Chromosomes contain controlling elements which can regulate the expression of genes during development, and what appears to be the high mutability of an unstable locus can be referred to changes involving the controller and not the structural gene itself. Controllers can change from an active to an inactive state and vice versa. A frequently observed effect of a controller is to repress the action of an associated gene, either partially or completely.

(b) Controllers may be autonomous in action, but frequently a controller effect is dependent on the presence of a further and specific controller unit in the same nucleus. Thus controllers themselves appear to be organized into systems whose members interact with each other.

(c) There is no evidence for a one-to-one functional relation between a gene and a controller. Some controllers appear to be transposable and impose their particular effects on those loci with which they happen to become associated.

(d) Although direct proof is lacking, McClintock considers it probable that controllers are associated with heterochromatin.

The bizarre developmental phenomena associated with unstable loci, like those connected with V-type position effects (see p. 77), are thus interpretable in terms of heterochromatin disarrangements affecting a general chromosome process that is antecedant to the control of individual gene action and distinct from it. If this is true it would appear that the activation of a gene necessitates two distinct sets of factors. One set, of a general nature, operates at the chromosome level, the other acts in a highly specific way at the individual gene level. Thus hetero-chromatinization functions as a non-specific repressor which blocks uncoiling, one of the steps in gene potentiation which precedes definitive gene activation. Alternatively heterochromatin can be viewed not so much as a mechanism to induce repression but rather as a mechanism to augment and maintain a repressed state. In individuals heterozygous for the X-linked enzyme glucose 6-phosphate dehydrogenase it is possible by cloning to obtain a cell culture in which one allele is inactive in all cells. During the log phase of growth, however, over 50 per cent of these cells show no sex chromatin body. Thus even when decondensed the X still remains inactive. Condensation is thus presumably secondary to inactivation.

There is a further sense in which the arrangement of genes within the eucaryote system appears to present a distinctive pattern of distribution. Rick (1971) has recently drawn attention to the interesting fact that in cases where linkage maps are sufficiently well known it can be seen that:

(a) in terms of pachytene euchromatin length the number of loci departs significantly from randomness. In tomato this deviation is determined principally by a concentration of loci on chromosome 11 and a disparity of loci on chromosomes 5 and 12. In maize chromosome 8 is deficient and chromosome 10 loaded with loci (Fig. 48);

(b) within chromosomes too genes are concentrated in closely clustered groups. This is particularly evident near the centromere (Fig. 49) where, interestingly, there is commonly a concentration of highly reiterated DNA. High gene density is correlated with a low rate of recombination. In addition the pattern of gene clustering is quite different in the nucleolar organizing chromosome and significantly the recombination pattern in this chromosome is different from that of other chromosomes in the same complement.

The information contained in the hereditary hierarchy is, of course, used to regulate the development of the phenotype of an organism. The term phenotype is, however, a subtle one. It applies to and embodies any feature of an organism one cares to nominate. Thus the chromosome

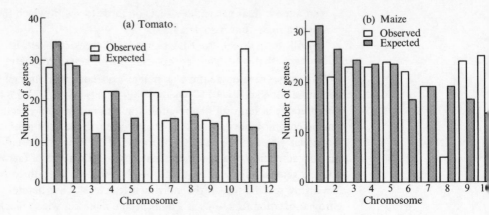

FIG. 48. Frequency distributions of nonallelic genes among the chromosomes of (a) tomato and (b) maize. Expected numbers are based on the proportional length of euchromatin in the pachytene chromosomes (after Rick (1971)).

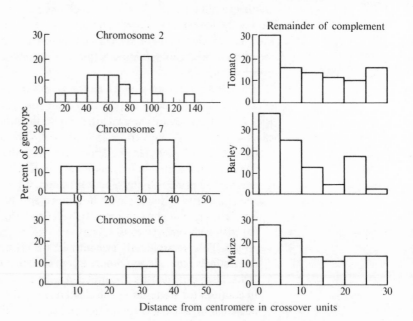

FIG. 49. The relationship between the nucleolar organizing element and the distribution of genes in the complements of tomato, barley, and maize. The distributions for the non-nucleolar chromosomes frequencies were pooled for those chromosomes mapped to at least thirty units. The number 6 nucleolar organizing element of barley has been omitted because only four of its genes have been mapped (after Rick, (1971)).

itself has a phenotypic aspect which is affected by both genetic and non-genetic factors. Its phenotype, however, is as close materially to the genotype as anything can possibly be, for the materials from which the genotype is constructed are included among those of which the chromosomes are made. Meiosis, of course, is an expression of chromosome phenotype, as too is fertilization. And mutations are known which influence the component stages of meiosis and the capacity of the egg to develop with or without fertilization. That is, the genes that are recombined and segregated at meiosis and then combined at fertilization include those responsible for the control of meiosis and fertilization. The chromosomes which carry the genes are thus subject to the control of the genes they carry and where there is a basis for inherited variation there is also a basis for selection and evolution. In consequence the state of the chromosome hierarchy is itself something compiled by selection. The chromosome hierarchy is thus also a dynamic hierarchy.

The living world provides many examples of different ways of achieving much the same end. This applies to the evolution of the chromosome hierarchy too. Thus, although many genetic systems are superficially very different structurally, they are essentially similar functionally. We saw earlier how the genetic consequences of subsexual apomixis are virtually indistinguishable from those of interchange hybridity combined with inbreeding (see p. 117). Similarly the meiotic consequences of inversion hybridity are similar in many ways to those resulting from a change in chiasma localization (see p. 106). Why then should different types follow different evolutionary pathways in order to arrive at essentially the same adaptive goal? Clearly the directions of successful change must be limited by the nature of the original genetic system. This, in turn, is defined by both chromosome structure and genotypic properties. The structure of the hierarchy thus limits the changes that successfully can occur in it. It is to these limits that we finally turn.

5.1. The prospects for interchange hybridity

Gametic fertility in interchange hybrids depends on the regular:

(a) formation of the maximum multiple association (pairing);
(b) maintenance of the maximum multiple association (chiasma frequency);
(c) restriction of crossing-over to the pairing segments (chiasma localization); and
(d) alternative orientation of the multiple.

These limitations are of two kinds, structural and genotypic.

5.1.1. Structural limitations

(a) *Multiple formation.* The minimum number of chiasmata required to maintain a multiple association ((b) above) is one less than the number of chromosomes in the multiple i.e. $1 + 2i$ where i is the number of independent interchanges. For fertility, therefore, there must be at least this number of pairing segments ((c) above). Pairing segments actually represent regions of two kinds. First, whole arms not involved in interchange (type (*i*)) and second, segments distal to the interchange points in the arms involved (type (*ii*)). In the case of metacentric chromosomes, the joint frequency of these two kinds of pairing segments is constant and equal to the number of chromosomes in the multiple. The relative frequency of these two kinds of pairing segment, on the other hand, depends on the structure of the multiple. But more important is the fact that the absolute number of type (*i*) pairing segments depends on whether the chromosomes involved are one or two armed. Thus a cross of four produced by a single interchange between telocentric chromosomes will have only type (*ii*) pairing segments and only two pairing segments in all. Therefore, since a minimum number of three chiasmata is required to maintain an association of four chromosomes, at least one of these must occur in an interstitial segment. Consequently fertility is unavoidably reduced by 50 per cent by each interchange of this type. Further unbranched multiples of six or more chromosomes cannot arise unless the chromosomes involve in double interchange are two armed.

Successful interchange is thus prohibited by telocentricity and rendered progressively less likely as arm ratios increase. In such cases interchange hybridity would have to await pericentric inversion or a comparable change capable of producing isobrachial chromosomes.

(b) *Multiple orientation*. Alternate orientation in rings means that each centromere is co-oriented in relation to the two centromeres on either side of it, and in large rings this co-orientation involves an increasing number of non homologus centromeres. Presumably such co-orientation reflects and require each centromere to be in equal communication with those on each side of it Centromeres communicate by movements, the effects of which are physicall transmitted via the attachments formed, usually by chiasmata. Equal communication is facilitated, therefore, by equal distances between centromere pairs in the ring. Thus, other things being equal, alternate orientation is expected to predominate when the chromosomes in the interchange comple: are equal in size and isobrachial. In general, this situation will obtain after th interchanges have arisen only if it exists initially and the interchanged segme are of equal length. Further, it is not only the number of pairing segments which is important but also their length, for they are required to show consistent chiasma formation. This means some preference with regard to the absolute sites of breakage. To some extent, therefore, the structural properti which favour multiple formation with crossing-over only in pairing segments also favour the alternate orientation of the multiple.

We see these preferences elegantly demonstrated in the experimentally induced interchange heterozygotes of diploid *Chrysanthemum* ($2n = 18$).

Chrysanthemum has a highly symmetrical karyotype consisting entirely of metacentric chromosomes. Rana (1965) obtained 87 interchange heterozygotes, involving from one to three interchanges, after irradiating dry seed with 15 kR of X-rays and the largest multiple observed was one involving 8 chromosomes. Plants carrying three interchanges were inter-crossed among themselves and the F_1 seeds thus obtained were irradiated with an X-ray dose of 10 kR. In the following generation, plants carrying 4 or 5 interchanges and forming rings or chains of as many as 10 or even 12 chromosomes were found (Table 27). Notice that in no case was

Table 27
Induced interchanges in Chrysanthemum

No. interchanges	Type of multiple	No. plants analysed	Frequency of disjunction (%)
1	1.IV	25	90·2
2	2.IV	15	87·2
	1.VI	15	90·0
3	3.IV	2	74·1
	1.VI + 1.IV	5	85·2
	1.VIII	5	89·9
4	2.VI	2	76·5
	2.VIII + 1.IV	3	76·3
	1.X	4	79·2
5	2.VIII + 2.VI	3	72·0
	1.X + 1.IV	4	69·3
	1.XII	3	74·8

Note. After Rana (1965).

fertility less than 70 per cent. Significantly the chromosome segments involved in the interchanges tended to be of equal size so that the symmetry of the karyotype was not disturbed. Also the interchange multiples formed mostly ring configurations and these were terminally associated, since even in bivalents of *Chrysanthemum* chiasmata are distally localized. Similarly in *Tradescantia palludosa*, with equivalent structural properties, Watanabe (1962) succeeded in inducing a maximum ring of XII (see p. 115) by X-irradiation.

5.1.2. Genotypic limitations

Of course the morphological aspects of multiple morphology which are expected to enhance its orientation behaviour are not necessarily sufficient to ensure it. Equally, morphologies which do not appear to facilitate alternate orientation are not sufficient to prohibit it. Within limits, both may be overriden by genotypic considerations. For example, the equidistant relationship between centromeres referred to above will not obtain, even when the chromosomes are equal in size and isobrachial, if the chiasmata lie at various and variable distances from the centromeres. Thus interchange hybridity is more likely to succeed in forms with distally localized chiasmata or chiasma terminalization. Chiasma frequency, chiasma localization, and chiasma terminalization are all known to be under genotypic control.

Flexibility of the multiple is also important because alternate orientation requires one twist for each interchange. Chiasma terminalization is important in this connection too since this will increase the flexibility of the multiple. However, flexibility and regular multiple formation make opposite demands with regard to chiasma frequency. Regular formation is favoured by high chiasma frequency but this reduces flexibility. Indeed, the orientation behaviour of a multiple is known to vary with the number of chiasmata in it. Thus to accommodate regularity of both formation and flexibility, the chiasma frequency per nucleus, and even per arm, must be regulated within very narrow limits.

From the above discussion it is clear that a system of permanent interchange hybridity is not easily evolved especially if the properties of the initial system are not propitious. Further, species which have developed such a system must have been under considerable selection pressure. But it is equally clear that no weight of selection can lead to the establishment of interchange hybridity unless the original type is characterized by a number of particular features with regard to both chromosome structure and various genotypically controlled aspects of chromosome behaviour. Experimental studies by Lawrence (1958) and by Sun and Rees (1967) have shown that selection is certainly effective in producing increases in the disjunction frequency of interchange configurations. The increases observed proved to be dependent on a reduction in chiasma frequency. Significantly, interchange heterozygosity in nature is commonly established and maintained in conditions where inbreeding is imposed on a normally outbreeding species. In such circumstances inbreeding would be expected to lower chiasma frequency so that this too would act as a facilitation force.

5.2. The prospects for polyploids
5.2.1. Autopolyploids

Gametic sterility in autopolyploids is determined mainly by the erratic and irregular segregation of univalents and of uneven multivalent associations. Consequently little can be done to improve sexual fertility in uneven polyploids and most of them are compelled to adopt vegetative or apomictic modes of reproduction.

In even polyploids, on the other hand, the sterilizing effects of uneven configurations can be avoided if their frequency can be reduced. In a tetraploid, for example, this may involve increased bivalent formation or increased quadrivalent formation. As far as the first solution is concerned there are two ways of realizing it.

(a) *Structural change.* A structural change within one member of a potential quadrivalent may reduce its chances of pairing with any of the other three. By inbreeding, this change could become homozygous, so raising it to the duplex level and increasing the prospects for bivalent formation. A modified autotetraploid of this type would have certain of the properties of an allopolyploid, including disomic inheritance for genes on the affected chromosomes. Indeed the apparent rarity of sexual autopolyploids in nature may be due to the accumulation of such structural changes, though it must be admitted that this kind of adjustment involves separate treatment for each chromosome type.

(b) *Genotypic change.* The minimum number of chiasmata required to produce an association of III + I is the same as that needed to give 2II, namely two. But the production of a trivalent requires a single chromosome to be involved in both of these exchanges. Thus bivalents could be formed in preference to uneven associations if the pattern of pairing, or exchange could be adjusted so that a given chromosome was limited to one chiasma. In general, chiasma interference seems to be effective, or more effective within, rather than between, chromosome arms. On this basis the above adjustment would be progressively more likely as the arm ratios of the chromosome increased. But, like genotypic control in general, it is expected to have a more uniform effect on the whole complement. Unlike structural adjustment, however, it gives random rather than preferred bivalents.

Replacing configurations of III + I with associations of IV cannot be achieved by internal structural change because this requires the retention of maximum homology. But this could result from selection for higher chiasma frequencies and, unlike selection for bivalents, its prospects are likely to increase as arm ratio decreases.

Selection for improved fertility in a colchicine-induced autotetraploid, *Lolium*, by Crowley and Rees (1968) was achieved at least in part by an increase in quadrivalent frequency. Moreover, the higher chiasma frequency of the quadrivalents in the selected material was largely accomplished by a redistribution of chiasmata rather than a change in over-all chiasma frequency. In *Lolium* the number of chiasmata in quadrivalents is low and, more important, chiasmata are located distally. Both these features are conducive to regular disjunction. In tetraploids where the chiasmata are more numerous and proximal rather than distal, diploidization may well be the only means of improving fertility.

2.2. *Allopolyploids*   Raw allopolyploids have problems similar to those of autopolyploids though they are less pronounced. In allopolyploids too, bivalent formation can be enhanced by increasing still further the degree of structural difference between homoeologues. In fact, a long paracentric inversion in maize has been found to increase preferential pairing in an allopolyploid produced by crossing autotetraploids of maize and teosinte.

Enhancement of bivalent formation by the genotypic intensification of preferential pairing has been described also. Thus pairing between homoeologues in hexaploid wheat is suppressed by a determinant on the long arm of chromosome V. This chromosome is thought to be a member of the B-genome contributed by *Aegilops speltoides* but homoeologous pairing does occur in *T. aestivum* × *Aegilops speltoides* hybrids. It would appear, therefore, that the critical mutation is recessive and that it occurred after the introduction of the B-genome into wheat.

An equivalent system appears to obtain in oats (Fig. 50), for here

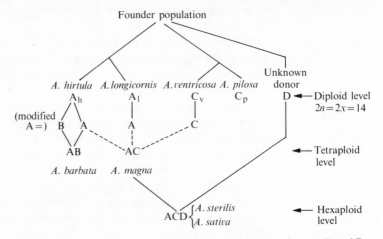

FIG. 50. Genome differentiation in *Avena*. The evolution of the A, C, and D genomes is assumed to reflect a phase of major re-patterning followed by relatively minor changes which give rise to the A and C subgroups. Polyploid convergence is then involved reducing the number of karyotypes to two at the tetraploid and one at the hexaploid level (after Rajhathy (1971)).

too hexaploids are strictly bivalent formers, and in *Rosa*. Thus in the pentaploid hybrid produced by crossing *R. canina* ($2n = 5x = 35$) and *R. rugosa* ($2n = 2x = 14$), using the former as female parent, 11–14 bivalents are formed. Even if all seven of the chromosomes derived from *rugosa* are involved in these bivalents, as appears likely, there is still pairing between 8–14 of the chromosomes derived from *canina*. In *R. canina* ($2n = 5x = 35 = $ AABCD) itself, however, these never pair (AII + BI + CI + DI) but their behaviour in this interspecific hybrid shows that the four types of genome in *R. canina* are not as different as the letters ABCD imply.

5.3. The prospects for apomixis

Female meiosis in apomicts is generally and in various ways different from that in sexual forms. To what extent are these differences a consequence of genotypic adjustment rather than structural control? Clearly, in those apomicts where the first division is normal, the modifications which do obtain can certainly be attributed to genotypic control. But where pairing fails and, consequently, the prospect of chiasma formation and co-orientation does not arise, the extent to which the initial anomaly depends on genotypic adjustment rather than structural non-homology is not immediately obvious. Further, the presence of a large number of univalents at first division in itself often results in spindle abnormalities and subsequent restitution—even in non-apomict forms—whatever the cause of the initial failure of pairing. Consequently, failure of pairing itself may be important in facilitating restitution and the suppression of first anaphase in apomicts.

In general, however, the genotypic constitution must be regarded as the crucial factor. Thus it is extremely unlikely that the three genomes in *Lumbricillus lineatus* ($2n = 3x$) are all different and share so little homoeology that no pairing can occur. Indeed it could be argued that this species is an autotriploid in which univalents are formed exclusively for purely genotypic reasons. Similarly the pattern of male meiosis in many apomictic plants shows that non-pairing on the female side is, for the most part, a genotypically controlled adaptation and not a structurally determined inevitability.

The evolution of apomixis depends, of course, not only on the circumvention of reduction but also on the capacity of the egg to develop without fertilization. The necessity for pre-adaptation in these respects is even greater and more obvious in the case of apomicts than in those considered earlier. Thus apomixis often obtains in organisms which are prohibited from reproducing sexually because of their chromosome constitution (e.g. uneven polyploids). Consequently, unless the capacity for apomixis exists prior to the origin of the sexual impediment it is not likely to arise at all. That is, species which are obliged to adopt apomixis are expected to arise from those which already include it facultatively in their reproductive repertoires.

Many biologists of the past and some of the present have seen or sought direction in both evolution and development. Indeed, interpretations based on recapitulation theories attempted to determine the direction of the former from the course of the latter. Of course, there are preferred pathways in both these expressions of genotype-environment interaction. But the canalization comes not from the attraction of a yet-to-be final form but from the information and organization which obtains initially and the contemporary forces to which it is subjected.

These forces are selective, influencing either the pattern of gene expression, as in development, or the prospects of genotype survival, as in evolution. Both development and evolution are driven not led.

The chromosomes constitute the principal basis of hereditary transmission and, consequently, changes in them are involved in both these processes. However, the nature of the chromosome changes, far from being recapitulative, is quite different in the two cases. Differentiation during development can be, and usually is, determined solely by differential gene action. Evolutionary divergence, on the other hand, relies on stable changes in the very nature of the genetic material itself and their differential propagation.

Bibliography

1. Introduction

(a) *General texts*

De Busk, A. G. (1968). *Molecular genetics.* Macmillan, New York.

Lewis, K. R. and John, B. (1970). *The organization of heredity.* Edward Arnold, London.

Ohno, S. (1970). *Evolution by gene duplication.* Springer-Verlag, Berlin.

Papazian, H. P. (1967). *Modern genetics.* Weidenfeld and Nicholson, London.

Stebbins, G. L. (1969). *The basis of progressive evolution.* University of North Carolina Press, Chapell Hill.

(b) *Specialist papers*

Bostock, C. (1971). Repetitious DNA. *Adv. Cell Biol.* 2, 153—223.

Britten, R. J. and Kohne, D. E. (1968). Repeated sequences in DNA. *Science* 161, 529—40.

—— Davidson, E. H. (1971). Repetitive and non-repetitive DNA sequences and a speculation on the origins of evolutionary novelty. *Quart. Rev. Biol.* 46, 111—33.

2. Chromosome architecture

(a) *General texts*

DuPraw, E. J. (1970). *DNA and chromosomes.* Holt, Rinehart, and Winston, New York.

John, B. and Lewis, K. R. (1972). *Somatic cell division.* Oxford Biology Readers No. 26, Oxford University Press.

———— (1973). *The meiotic mechanism.* Oxford Biology Readers No. 65. Oxford University Press, Oxford.

Ohno, S. (1970). *Evolution by gene duplication.* Springer-Verlag, Berlin.

Swanson, C. P., Mertz, T., and Young, W. J. (1967). *Cytogenetics.* Prentice—Hall, New Jersey.

Whitehouse, H. K. L. (1973). *The mechanism of heredity* (3rd edn). Edward Arnold, London.

(b) *Specialist papers*

Ashburner, M. (1970). The genetic analysis of puffing in polytene chromosomes of *Drosophila. Proc. R. Soc.* B176, 319—27.

Bachmann, K., Goin, O. B., and Goin, C. J. (1972). Nuclear DNA amounts in vertebrates. *Brookhaven Symp. Biol.* 23, 419—47.

Beçak, W., Beçak, M. L., Schreiber, G., Lavalle, D., and Amonum, F. O. (1970). Interspecific variability of DNA content in Amphibia. *Experientia* 26, 204–6.

Beçak, M. L., Denaro, L., and Beçak, W. (1970). Polyploidy and mechanisms of karyotypic diversification in Amphibia. *Cytogenetics* 9, 225–38.

Beermann, W. (1961). Ein Balbiani Ring als Locus einer Speicheldrusen-mutation. *Chromosoma* 12, 1–25.

—— (1972). Chromosomes and genes. *Res. Prob. Cell Diff.* 4, 1–33.

Bogart, J. P. and Wasserman, A. O. (1972). Diploid-polyploid cryptic species pairs: a possible clue to evolutionary polyploidisation in anuran amphibians. *Cytogenetics* 11, 7–24.

Braselton, J. P. (1971). The ultrastructure of the non-localised kinetochores of *Luzula* and *Cyperus*. *Chromosoma* 36, 88–99.

Britten, R. J. and Davidson, E. H. (1971). Repetitive and non-repetitive DNA sequences and a speculation on the origins of evolutionary novelty *Quart. Rev. Biol.* 46, 111–33.

Brown, S. W. (1966). Heterochromatin. *Science* 151, 417–25.

Callan, H. G. (1972). Replication of DNA in the chromosomes of eukaryotes. *Proc. R. Soc.* B181, 19–41.

Cassagnau, P. (1968). Sur la structure des chromosomes Salivaires de *Bilobella massoudi* Cassagnau (Collembola: Neamuridae). *Chromosoma* 24, 42–58.

—— (1971). Les chromosomes salivaries polytènes chez *Bilobella grassei* (Denis) (Collemboles: Neanuridae). *Chromosoma* 35, 57–83.

Cleveland, L. R. (1949). The whole life cycle of chromosomes and their coiling systems. *Trans. Am. phil. Soc.* 39, 1–100.

Comings, D. E. (1972). The structure and function of chromatin. *Adv. Human Genet.* 3, 237–431.

—— Okada, T. A. (1972) Holocentric chromosomes in *Oncopeltus:* Kinetochore plates are present in mitosis but absent in meiosis. *Chromosoma* 37, 177–92.

Gall, J. G., Cohen, E. H., and Polan, M. L., (1971). Repetitive DNA sequences in *Drosophila*. *Chromosoma* 33, 319–44.

Gassner, G. (1969). Synaptinemal complexes in the achiasmatic spermatogenesis of *Bolbe nigra* Giglio-Tos (Mantoidea). *Chromosoma* 26, 22–34.

Gay, H., Das, C. C., Forward, K., and Kaufmann, B. P. (1970). DNA content of mitotically active condensed chromosomes of *D. melanogaster*. *Chromosoma* 32, 213–23.

Gerstel, D. U. and Burns, J. A. (1967). Phenotypic and chromosomal abnormalities associated with the introduction of heterochromatin from *Nicotiana otophora* into *N. tabacum*. *Genetics* 56, 583–602.

Grell, R. F. (1967). Pairing at the chromosome level. *J. Cell Physiol.* Suppl. 1, 70, 1119–46.

Hennig, W., Hennig, I., and Stein, H. (1970). Repeated sequences in the DNA of *Drosophila* and their location in giant chromosomes. *Chromosoma* 32, 31–63.

Hotta, Y. and Stern, H. (1971). Analysis of DNA synthesis during meiotic prophase in *Lilium*. *J. Molec. Biol.* 55, 337–55.

Hsu, T. C., Cooper, J. E. K., Mace, M. L., and Brinkley, B. R. (1971). Arrangement of centromeres in mouse cells. *Chromosoma* 34, 73–87.

Hubermann, J. A. and Riggs, A. D. (1968). On the mechanism of DNA replication in mammalian chromosomes. *J. mol. Biol.* **32**, 327–41.

Inoué, S. (1964). Organization and function of the mitotic spindle. In *Primitive motile systems in cell biology* pp. 549–94. Academic Press, New York.

John B. and Lewis, K. R. (1965). *The meiotic system. Protoplasmatologia V1F1.* Springer-Verlag, Vienna.

Jones, R. N. and Rees, H. (1968). Nuclear DNA variation in *Allium. Heredity* **23**, 591–605.

Kavenoff, R. and Zimm, B. H. (1973). Chromosome-sized DNA molecules from *Drosophila. Chromosoma* **41**, 1–27.

Keyl, H.-G. (1965). A demonstrable local and geometric increase in the chromosomal DNA of *Chironomus. Experientia* 21, 191.

Masters, M. M. and Broda, P. (1971). Evidence for the bidirectional replication of the *E. Coli* chromosome. *Nature New Biol.* **232**, 137–40.

Nicklas, R. B. (1972). Mitosis. *Adv. Cell Biol.* **2**, 225–97.

Ohno, S. and Atkin, N. B. (1966). Comparative DNA values and chromosome complements of eight species of fishes. *Chromosoma* **18**, 455–66.

–– Wolf, U., and Atkin, N. B. (1968). Evolution from fish to mammals by gene duplication. *Hereditas* **59**, 169–87.

Pardue, M. L., Brown, D. D., and Birnstiel, M. L. (1973). Location of genes for 5S ribosomal RNA in *Xenopus laevis. Chromosoma* **42**, 191–203.

Peacock, W. J. and Miklos, G. G. (1973). Meiotic drive in *Drosophila*: new interpretations of the segregation distorter and sex chromosome systems. *Adv. Genet.* **17**, 361–409.

Perondini, A. L. P. and Dessen, E. M. (1969). Heterozygous puffs in *Sciara ocellaris. Genetics* **61**, Suppl., 251–60.

Prescott, D. E. (1970). The structure and replication of eukaryotic chromosomes. *Adv. Cell Biol.* **1**, 57–117.

Rees, H. (1972). DNA in higher plants. *Brookhaven Symp. Biol.* |23, 394–418.

–– Jones, R. N. (1972). The origin of the wide species variation in nuclear DNA content. *Int. Rev. Cytol.* **32**, 53–92.

Ribbert, D. and Bier, K. (1969). Multiple nucleoli and enhanced nucleolar activity of the insect ovary. *Chromosoma* **27**, 178–97.

Riley, R. (1966). Genetics and the regulation of meiotic chromosome behaviour. *Sci. Prog.* **54**, 193–207.

Sparrow, A. H., Price, H. J., and Underbrink, A. G. (1972). A survey of DNA content per cell and per chromosome of prokaryotic and eukaryotic organisms: Some evolutionary considerations. *Brookhaven Symp. Biol.* **23**, 451–93.

Tartof, K. D. (1973). Regulation of ribosomal RNA gene multiplicity in *Drosophila melanogaster. Genetics* **73**, 57–71.

Whitehouse, H. L. K. (1965). Crossing-over. *Sci. Prog.* **53**, 285–96.

Wolff, S. (1969). Strandedness of chromosomes. *Int. Rev. Cytol.* **25**, 279–96.

3. Epigenetic activities

(a) *General texts*

Bonner, J. (1965). *The molecular biology of development.* Clarendon Press, Oxford.

Davidson, E. H. (1969). *Gene activity in early development.* Academic Press, New York.
John, B. and Lewis, K. R. (1969). *Chromosome cycle. Protoplasmatologia VIB.* Springer-Verlag, Vienna.
Markert, C. L. and Ursprung, H. (1971). *Developmental genetics.* Prentice-Hall, New Jersey.

(b) *Specialist papers*
Baker, W. K. (1968). Position-effect variegation. *Adv. Genet.* **14**, 133–69.
Beermann, S. (1966). A quantitative study of chromatin dimunution in embryonic mitoses of *Cyclops* furcifer. *Genetics* **54**, 567–76.
Berendes, H. D. (1968). Factors involved in the expression of gene activity in polytene chromosomes. *Chromosoma* **24**, 418–37.
Bonner, J. and Huang, R. C. (1966). Histones as specific repressors of chromosomal RNA synthesis. In *Histones, their role in the transfer of genetic information,* pp. 18–33. Ciba Foundation Study No. 24. Churchill, London.
—— Dahms, M., Fambrough, D., Huang, R. C., Marushige, K. and Tuan, D. (1968). Composition and template activity of chromatin from various tissues. *Science* **159**, 47.
Bostock, C. J. and Prescott, D. M. (1972). Evidence of gene diminution during the formation of the macronucleus in the protozoan *Stylonchia. Proc. natn. Acad. Sci. U.S.A.* **69**, 139–42.
Boveri, T. (1892). Die Entstehing des Gegensatzes Zwischen den Geschlechtzellen und den somatischen Zellen bei *Ascaris megalocephala. Sber. Ges. Morph. Physiol. München.* **8**, 114–125.
Brink, R. A. (1964). Genetic repression in multicellular organisms. *Am. Nat.* **98**, 193–211.
Brown, S. W. and Nur, U. (1964). Heterochromatic chromosomes in the coccids. *Science* **145**, 130–6.
Callan, H. G. (1972). The organization of genetic units in chromosomes. *J. Cell Sci.* **2**, 1–7.
Carson, H. L. (1967). Selection for parthenogenesis in *D. mercatorum. Genetics* **55**, 157–71.
—— Hardy, D. E. Spieth, H. T., and Stone, W. S. (1970). The evolutionary biology of the Hawaiian Drosophilidea. In *Essays in evolution and genetics in honor of Th. Dobzhansky,* pp. 457–543. Appleton-Century-Crofts, New York.
Cattenach, B. M. (1961). A chemically induced variegated-type position effect in the mouse. *Z. VererbLehre* **92**, 165–82.
Crouse, H. V. and Keyl, H.-G. (1968). Extra-replications in the 'DNA-puffs' of *Sciara coprophila. Chromosoma* **25**, 357–64.
De Cunha, A. B., Pavan, C., Morgante, J. S., and Garrido, M. C. (1969). Studies on cytology and differentiation in Sciaridae II. DNA redundancy in salivary gland cells of *Hybosciara fragilis* (Diptera, Sciaridae). *Genetics* **61**, Suppl., 335–49.
Evans H. J., Ford, C. E., Lyon, M. F., and Gray, J. (1965). DNA replication and genetic expression in female mice with morphologically distinguishable X-chromosomes. *Nature* **206**, 900–3.
Gall, J. G. (1969). The genes for ribosomal RNA during oogenesis. *Genetics* **61**, Suppl., 121–32.
Hennig, W. and Meer, B. (1971). Reduced polyteny of ribosomal RNA

cistrons in giant chromosomes of *Drosophila hydei. Nature, New Biol.*
233, 70—2.

—— (1971). Lampenbürstenchromosomen. *Hanbuch Allgem. Pathologie,*
pp. 215—81.

Huang, R. C., Bonner, J., and Murrary, K. (1964). Physical and
biological properties of nucleohistones. *J. mol. Biol.* 8, 54—64.

John, B. and Lewis, K. R. (1968). *Chromosome complement. Proto-
plasmatologia VIA.* Springer-Verlag, Vienna.

Kaulenas, M. S. and Fairbairn, D. (1966). Ribonuclease-stable polysomes
in the egg of *Ascaris lumbricoides. Devl. Biol.* 14, 481—94.

Kroeger, H. (1964). Zellphysiologische Mechanismen bei der Regulation
von Genaktivitaten in den Riesenchromosomen von *Chironomus
thummi. Chromosoma* 15, 36—70.

Kunz, W., Trepte, H.-H., and Bier, K. (1970). On the function of the
germ line chromosomes in the oogenesis of *Wachtiella persicariae*
(Cecidomyidae). *Chromosoma* 30, 180—92.

Lezzi, M. and Robert, M. (1972). Chromosomes isolated from unfixed
salivary glands of *Chironomus. Cell Diff.* 4, 35—57.

Lima-de-Faria, A. and Moses, M. J. (1966). Ultrastructure and cyto-
chemistry of metabolic DNA in *Tipula. J. Cell Biol.* 30, 177—92.

Lyon, M. F. (1962). Sex chromatin and gene action in the mammalian
X-chromosome. *Am. J. hum. Genet.* 14, 135—48.

—— (1968). Chromosomal and subchromosomal inactivation. *A. Rev.
Genet.* 2, 31—52.

—— (1972). X-chromosome inactivation and developmental patterns in
mammals. *Biol. Rev.* 47, 1—35.

Macgregor, H. C. (1972). The nucleolus and its genes in amphibian
oogenesis. *Biol. Rev.* 47, 177—210.

Mancino. G., Nardi, I., and Barsacchi, G. (1970). Spontaneous aberrations
in lampbrush chromosome XI from a specimen of *Titurus vulgaris
meridionalis* (Amphibia, Urodela). *Cytogenetics* 9, 260—71.

Marushige, K. and Bonner, J. (1966). Tempate properties of liver
chromatin. *J. mol. Biol.* 15, 160—74.

Miller, O. L. and Beatty, B. R. (1969). Portrait of a gene. *J. Cell Physiol.*
74, Suppl., 225—32.

Nur, U. (1966). Non-replication of heterochromatic chromomes in a
mealy bug, *Planococus citri* (Coccoidea, Homoptera). *Chromosoma*
19, 439—48.

—— (1968). Synapsis and crossing-over within a paracentric inversion in
the grasshopper *Camnula pellucida. Chromosoma* 25, 198—214.

Ohno, S. and Cattanach, B. M. (1962). Cytological studies of an
X-autosome translocation in *Mus musculus. Cytogenetics* 1, 129—40.

Painter, T. S. (1966). The rqle of the E-chromosomes in Cecidomyidae.
Proc. natn. Acad. Sci. U.S.A. 56, 853—5.

—— and Biesele, J. J. (1966). Endomitosis and polyribosome formation.
Proc. natn. Acad. Sci. U.S.A. 56, 1920—5.

Paul, J., Gilmor, R. S., Thomson, H., Threlfall, G., and Khol, D. (1970).
Organ-specific gene masking in mammalian chromosomes. *Proc. R.
Soc.* B176, 277—85.

Perondini, A. L. P. and Dessen, E. M. (1969). Heterozygous puffs in
Sciara ocellaris. Genetics 61 Suppl., 251—60.

Prokofyeva-Belgovskaya, A. A. (1947). Heterochromatinisation as a change of chromosome cycle. *J. Genet.* **48**, 80–98.

Stedman, E. and Stedman, E. (1947). The chemical nature and functions of the components of cell nuclei. *Cold Spring Harb. Symp. Quant. Biol.* **12**, 224–36.

Steffensen, D. M. and Wimber, D. E. (1971). Localisation of tRNA genes in the salivary chromosomes of *Drosophila* by RNA:DNA hybridisation. *Genetics* **69**, 163–78.

Swift, H. (1961). Nuclear physiology and differentiation. *Genetics* **61**, Suppl., 439–61.

Tomkins, G. M. and Martin, D. W. (1970). Hormones and genes expression. *Adv. Human Genet.* **14**, 91–106.

Wallace, H., Murray, J. and Langridge, W. H. R. (1971). Alternative model for gene amplification. *Nature, New Biol.* **230**, 201–3.

Whitten, J. M. (1964). Giant polytene chromosomes within hypodermal cells of the developing foot pads of dipteran pupae. *Science* **143**, 1437.

4. Phylogenetic functions

(a) *General texts*

Dobzhansky, Th. (1970). *Genetics of the evolutionary process.* Columbia University Press, New York.

Darlington, C. D. (1958). *Evolution of genetic systems* (2nd edn). Oliver and Boyd, Edinburgh.

Grant, V. (1971). *Plant speciation.* Columbia University Press, New York.

John, B. and Lewis, K. R. (1968). *Chromosome complement. Proto-plasmatologia VIA.* Springer-Verlag, Vienna.

Mayr, E. (1971). *Population, species and evolution.* Harvard University Press, Massachusetts.

Stebbins, G. L. (1971). *Chromosmal evolution in higher plants.* Edward Arnold, London.

White, M. J. D. (1973). *Animal cytology and evolution* (3rd edn). Cambridge University Press.

(b) *Specialist papers*

Carson H. L. (1967). Selection for parthenogenesis in *Drosophila mercatorum. Genetics* **55**, 157–71.

Craddock, E. (1970). Chromosome number variation in a stick insect *Didymuria violescens* (Leach). *Science* **167**, 1380.

Darlington, C. D. (1956). Natural populations and the breakdown of classical genetics. *Proc. R. Soc.* B**145**, 350–64.

Dobzhansky, Th. (1961). On the dynamics of chromosomal polymorphism in *Drosophila.* In *Insect polymorphism,* pp. 36–42. Royal Entomological Society, London.

–– (1972). Species of *Drosophila. Science* **177**, 664–9.

Enrendorfer, F. (1965). Dispersal mechanisms, genetic systems and colonising abilities in some flowering plant families. *The genetics of colonising species,* pp. 331–51. Academic Press, New York.

Ford, C. E. and Hamerton, J. L. (1970). Chromosome polymorphism in the common shrew, *Sorex araneus. Symp. zool. Soc. Lond.* **26**, 223–36.

Grant, V. (1958). The regulation of recombination in plants. *Cold Spring Harb. Symp. quant. Biol.* **23**, 337–63.

Gropp, A., Winking, Zech, L., and Miller, H. (1972). Robertsonian chromosomal variation and identification of metacentric chromosomes in feral mice. *Chromosoma* **39**, 265–88.

Hewitt, G. M. and John, B. (1972). Inter-population sex chromosome polymorphism in the grasshopper *Podisma pedestris* II Population parameters. *Chromosoma* **37**, 23–42.

Jaylet, A. (1971). Creation d'une lignee homozygote pour une translocation reciproque chez l'Amphibien *Pleurodeles waltlii*. *Chromosoma* **34**, 383–423.

John, B. and Lewis, K. R. (1966). Chromosome variability and geographic distribution in insects. *Science* **152**, 711–21.

Koch, P., Pijnacker, L. P., and Kreke, J. (1972). DNA reduplication during meiotic prophase in the oocytes of *Carusius morosus* Br. (Insecta, Cheleutoptera). *Chromosoma* **36**, 313–21.

Lecher, P. (1967). Cytogenetique de l'hybridisation experimentale et naturelle chez l'isopode *Jaera* (*albifrons*) *syei* Bocquet. *Archs Zool., Exp. Gen.* **108**, 633–98.

Levin, D. A., Howland, G. P., and Steiner, E. (1972). Protein polymorphism and genetic heterozygosity in a population of the permanent translocation heterozygote, *Oenothera biennis*. *Proc. natn. Acad. Sci. U.S.A.* **69**, 1475–77.

Lewis, H. (1966). Speciation in flowering plants. *Science* **152**, 167–72.

Lowe, Ch., Wright, J. W., Cole, Ch. J., and Bezy, R. L. (1970). Natural hybridisation between the teiid lizards *Cnemidophorus sonorae* (parthenogenetic) and *Cnemidophorus tigris* (bisexual). *System. Zool.* **19**, 115–27.

Martin, J. (1967). Meiosis in inversion heterozygotes in Chironomidae. *Can. J. Genet. Cytol.* **9**, 255–68.

Nadler, C. F., Hoffmann, R. S., and Greer, K. R. (1971). Chromosomal divergence during evolution of ground squirrel populations (Rodentia: Spermophilus). *System. Zool.* **20**, 298–305.

Nur, U. (1968). Synapsis and crossing over within a paracentric inversion in the grasshopper *Camnula pellucida*. *Chromosoma* **25**, 198–214.

Ohno, S., Jainchill, J., and Stenius, C. (1963). The creeping vole (*Microtus oregoni*) as a gonosomic mosaic. The OX/XY constitution of the male. *Cytogenetic* **2**, 232–9.

Patton, J. L. (1973). Patterns of geographic variation in karyotype in the pocket gopher, *Thomomys bottae* (Eydoux and Garrais). *Evolution* **26**, 574–86.

—— Dingman, R. E. (1969). Chromosome studies of pocket gophers, genus *Thomomys* II. Variation in *T. bottae* in the American southwest. *Cytogenetics* **9**, 139–51.

Saez, F. A. and Mosquera, G. P. (1971). Cytogenetica del Genero *Dichroplus* (Orthoptera: Acrididea). *Rec. Adel. en Biol. 5° Congr. Argent de Ciencias Biol.*, p. 111–120.

Smith, S. G. (1966). Natural hybridisation in the coccinellid genus *Chilocorus*. *Chromosoma* **18**, 380–406.

Southern, D. I. (1967). Chiasma distribution in truxaline grasshoppers. *Chromosoma* **22**, 164–91.

Suomalainen, E. (1969). Evolution in parthenogenetic circulionidae. *Evol. Biol.* **3**, 261–96.

Tettenborn, U. and Gropp, A. (1970). Meiotic nondisjunction in mice and mouse hybrids. *Cytogenetics* **9**, 272–83.

Thaeler, Ch. S. (1968). Karyotypes of sixteen populations of the
Thomomys talpoides complex of pocket gophers (Rodentia-Geomyidae).
Chromosoma 25, 172–83.

Uzell, T. M. (1963). Natural triploids in salamanders related to
Ambystoma jeffersonianum. Science 139, 113–14.

Wahrman, J., Goitein, R., and Nevo, E. (1969). Geographic variation of
chromosome forms in *Spalax*, a subterranean mammal of restricted
mobility. In *Comparative mammalian cytogenetics*, pp. 30–48.
Springer-Verlag, New York.

Wet, de J. M. J. (1971). Reversible tetraploidy as an evolutionary
mechanism. *Evolution* 25, 545–8.

White, M. J. D. (1969). Chromosomal rearrangements and speciation in
animals. *A. Rev. Genet.* 3, 75–98.

—— (1970). Heterozygosity and genetic polymorphism in parthenogenetic
animals. *Evol. Biol.*, Suppl., 237–62.

Yosida, T. H. (1973). Evolution of karyotypes and differentiation in
thirteen *Rattus* species. *Chromosoma* 40, 285–98.

—— Tsuchiya, K. and Moriwaki, K. (1971). Karyotypic differences of
black rats, *Rattus rattus* collected in various localities of East and
S.E. Asia and Oceania. *Chromosoma* 33, 252–67.

Zohary, D. (1965). Coloniser species in the wheat group. In *Genetics of
colonising species*, pp. 403–19 (ed. H. G. Baker and G. L. Stebbins)
pp. 403–19.

5. Conclusion

(a) *General texts*

Lewis, K. R. and John, B. (1963). *Chromosome marker.* Churchill,
London.

(b) *Specialist papers*

Crowley, J. G. and Rees, H. (1968). Fertility and selection in tetraploid
Lolium. Chromosoma 24, 300–8.

Fincham, J. R. S. (1970). The regulation of gene mutation in plants. *Proc.
R. Soc.* B176, 295–302.

Horowitz, H. N. (1965). The evolution of biochemical synthesis—restrospect
and prospect. In *Evolving genes and proteins*, pp. 15–23. Academic
Press, New York.

Lawrence, C. W. (1958). Genotypic control of chromosome behaviour in
rye VI. Selection for disjunction frequency. *Heredity* 12, 127–31.

McClintock, B. (1965). The control of gene action in maize. In 'Genetic
control of differentiation'. *Brookhaven Symp. Biol.* 18, 162–84.

Rajhathy, T. (1971). The alloploid model in *Avena. Stadler Symp.* 3,
71–87,

Rana, R. S. (1965). Induced interchange heterozygosity in diploid
Chrysanthemum. Chromosoma 16, 477–85.

Rees, H. (1961). Genotypic control of chromosome form and behaviour.
Bot. Rev. 27, 288–318.

Rhoades, M. M. (1936). The effect of varying gene dosage on aleurone
colour in maize. *J. Genet.* 33, 347–54.

Rick, C. M. (1971). Some cytogenetic features of the genome in diploid
plant species. *Stadler Symposium*, 3, 153–74.

Sand, S. A. (1965). Position effects and the problems of coding a programme for development. *Am. Nat.* **99**, 33–45.

Sun, S. and Rees, H. (1967). Genotypic control of chromosome behaviour in rye IX. The effect of selection on the disjunction frequency of interchange associations. *Heredity* **22**, 249–54.

Watanabe, H. (1962). An X-ray induced strain of ring of twelve in *Tradescantia paludosa. Nature* **193**, 603.

Subject Index

166   *Subject Index*

Species Index